CAMBRIDGE COUNTY GEOGRAPHIES

SCOTLAND

General Editor: W. Murison, M.A.

MIDLOTHIAN

Cambridge County Geographies

MIDLOTHIAN

by

ALEX. McCALLUM, M.A., LL.B.

With Maps, Diagrams and Illustrations

Cambridge :
at the University Press
1912

CAMBRIDGE UNIVERSITY PRESS
Cambridge, New York, Melbourne, Madrid, Cape Town,
Singapore, São Paulo, Delhi, Mexico City

Cambridge University Press
The Edinburgh Building, Cambridge CB2 8RU, UK

Published in the United States of America by Cambridge University Press, New York

www.cambridge.org
Information on this title: www.cambridge.org/9781107620810

First published 1912
First paperback edition 2013

A catalogue record for this publication is available from the British Library

ISBN 978-1-107-62081-0 Paperback

CONTENTS

CONTENTS

ILLUSTRATIONS

MAPS

The illustrations on pp. 95, 123, 125, 126, 137, 147, 150, and 154 are reproduced from photographs by Messrs W. Ritchie and

Son; those on pp. 3, 9, 12, 16, 18, 21, 39, 41, 72, 78, 82, 84, 93, 96, 98, 100, 102, 112, 114, 116, 117, 119, 120, 121, 128, 129, 131, 132, 135, 136, 139, 140, 143, 145, 146, 148, 151, 169, 171, 178, 185, 187, 188, 190, 191, 193, 194, 196, 197, 199, 201, and 203 are from photographs by Messrs J. Valentine and Sons; those on pp. 91, 165, and 166 are reproduced from Cowan's *John Knox* by courtesy of Messrs G. P. Putnam's Sons; that on p. 107 is reproduced by kind permission of the Society of Antiquaries of Scotland; those on pp. 172, 173, 176, 177, 179, 181, and 182 are from photographs by Messrs T. and R. Annan, and that on p. 174 from a photograph by Mr Emery Walker.

1. County and Shire. Midlothian.

For purposes of administration Scotland is divided
into portions known as Shires or Counties. " Shire " is
usually said to be cognate with " share " and to mean
division, but this derivation is now disputed : a "county"
is the district which was at one time under the jurisdiction
of a Count. The division goes back as far as we have
written record, although the number of shires has varied
in course of time and the boundaries have not always
remained the same. The official who represented the
King's authority in the shire was known as the Shire-
reeve or Sheriff ; and sheriffdoms were modified in
number and in area from time to time as was found
convenient. Thus in 1305 there were at least twenty-
five, while at present there are thirty-three.

Probably, the sheriffdom of Edinburgh originally
extended over the whole of the Lothians ; later it was
defined as " a district extending from Colbrandspath (now
Cockburnspath) or Edgebucklin Brae on the east to the
water of Avon on the west " ; by limitations made at
various times it was reduced till it coincided with the
county as now defined ; and again, since 1870 the
Sheriff of Edinburgh exercises jurisdiction over Midlothian,
Linlithgow, Haddington, and Peebles.

The district of Lothian varied in extent from time to time, the name being vaguely applied to the low country south of the Forth as far as the Tweed. The significance of the name is uncertain. It may be derived from the Gaelic word for marsh or mire or alluvial land ; or it may be connected with the Saxon word for the people, *lèoda*, or that for a chief, *lèod*, in which case the term may have been applied to distinguish the Saxon inhabitants of the region from the Celtic tribes among whom they had settled. Whatever its origin the name has had several forms ; the Gaels called it Lethead ; the Saxons, Lothene ; while its Latin form was usually Lodoneia. Colloquially it was pronounced Lowden, as in R. L. Stevenson's poem, " A Lowden Sabbath Morn."

The name is now restricted to the counties of Linlithgow, Edinburgh, and Haddington, which are known respectively as West, Mid, and East Lothian. Ecclesiastically, the Synod of Lothian and Tweeddale corresponds roughly to the three civil districts named with the addition of Peeblesshire and part of Lanarkshire.

2. General Characteristics. Position and Natural Conditions.

Forming part of the southern shore of the Firth of Forth, containing tracts of the most fertile land in the kingdom and extensive areas of hill country well adapted for grazing, and having within its borders a rich coalfield, Midlothian is at once a maritime, an agricultural,

Edinburgh from the Castle

a pastoral, and an industrial region. With the capital of the country for its chief town, it may be regarded as the metropolitan county of Scotland, having a share in the legal, administrative, and educational functions of the metropolis.

The harbours of Leith and Granton are convenient gateways to the busy agricultural and manufacturing district round about them; fleets of steamers pass to and fro carrying to continental ports coal and the varied manufactures of the hinterland and bringing back multifarious cargoes for distribution throughout the country; the waters of the Firth and of the North Sea beyond yield an abundant harvest to the trawlers of Granton and the fishermen of Newhaven and Fisherrow; the collieries provide fuel for the household fires of the city and power for many factories and engineering works; the farmers of the Lothians are famous the world over for their skill and success in all the arts of husbandry; and the quiet walks of the Pentlands and the Moorfoots afford sustenance to numerous flocks of sheep.

Midlothian occupies the central eastern part of the belt of lowland which lies between the two great tablelands of Scotland—the Highlands and the Southern Uplands.

Edinburgh, the Heart of Midlothian, is in latitude 55° 57′ 23″ north and longitude 3° 10′ 30″ west. The same parallel of latitude passes near or through Copenhagen in Denmark, Moscow in Russia, the peninsula of Kamchatka in the east of Asia, Prince Rupert, the new port in Vancouver, and the inhospitable region of Labrador;

and a striking contrast may accordingly be instituted between the equable conditions prevailing in our district and those of the more or less favoured regions named. It is interesting also to notice that Edinburgh is really a little further west than Liverpool or Bristol, though these are situated on the opposite coast of the country.

The position of Midlothian on the sheltered waters of the Forth, with the safe anchorage of Leith Roads just off its shores, would in any case have given it a certain importance. Conveniently placed for trade with the Continent and having within its bounds the historic capital of the country, the county has had its importance greatly enhanced; while the advantages of a fertile soil, skilfully and carefully farmed, as well as a rich coalfield, have added still further to its wealth and greatness.

3. Size. Shape. Boundaries.

Measured from west to east, the greatest length of the county is 36 miles; its breadth from north-west to south-east is 24 miles; and its area is 370 square miles or 236,595 acres. As compared with other counties of Scotland, it comes twenty-second in magnitude of land area. Out of Inverness, the largest county, eleven and a quarter Midlothians might be carved; whereas it is about seven and a half times bigger than Clackmannan, smallest of Scottish shires.

In shape it resembles an Australian boomerang, with its convex side to the north and the ends turned to the south-east and south-west respectively.

It is bounded on the north by the Firth of Forth ; on the west by Linlithgowshire ; on the south by the counties of Lanark, Peebles, and Selkirk ; and on the east by Roxburghshire, Berwickshire and East Lothian.

The boundary with West Lothian is marked out by the course of the Almond and its tributary the Breich Water ; the southern limit is determined by the ranges of the Southern Pentlands and the Moorfoots ; while the eastern follows the line of the Moorfoots in its southern portion, Brothershiels and Dean Burns in the middle, and a variously named ridge of high ground in the northern.

4. Surface and General Features.

From the coast of the Firth of Forth in the north there is a gradual rise inland and southward towards the Pentlands and the Moorfoot Hills. The long slope between the hills and the sea may be described as a tilted plain dipping to the north. To the east this plain is terminated by the long ridge already mentioned—Roman Camp Hill—that separates Mid and East Lothian ; to the west it is continued into Linlithgowshire and the Carse of Stirling and Falkirk. Through the midst of the plain in Midlothian runs, from south-west to north-east, the ridge of the Pentland and the Braid Hills, which thus divides it into two basins. Apart from Arthur's Seat and the Calton Hill the eastern slope is broken only by gentle undulations ; but the western portion is more irregular. For seven miles west of Edinburgh the plain is fairly broad and continuous, but otherwise there are few level spaces of any considerable area. Rather has the surface the appearance of long waves, rising in successive ridges which run nearly east and west ; and even this regular alternation of ridge and valley is broken by the presence here and there of eminences which rise sharply out of the plain.

Thus it will be seen that, though there is nothing of the majestic or grand in the scenery of the county, there is yet abundance of pleasing variety. The peculiarly characteristic feature of the district—the occurrence of steep crags with gentle slopes tailing off from them— adds a strikingly picturesque element to the landscape. Finely wooded parks, deep and narrow glens, diversify the surface, and alternate with the fertile, highly cultivated farmlands of the northern plains ; the steep slopes of the Pentlands and the rounded rolling heights of the Moorfoots shut in the prospect to the south ; while the waters of the Forth, with the hills of Fife, the Ochils, and the Grampians rising in succession beyond, complete a scene of rare variety and beauty.

Three miles south of Edinburgh the Pentland Hills rise abruptly out of the plain and extend in a south-west direction through the middle of the county, continuing into the neighbouring shires of Peebles and Lanark. Their length is about sixteen miles, their breadth from four to six. They do not form a continuous chain, being cut into by many cross valleys, some of which afford passage from the one side of the hills to the other. Thus, about the middle of the range, through the Cauldstane Slap between the East and the West Cairn Hill, an old drove road leads from Lothian into Tweeddale and connects two main routes that pass along either flank of the hills—the one on the north being the Edinburgh and Lanark, and that to the south the Edinburgh and Dumfries road. Another of these old drove roads led from Currie by the Kirk Loan and the Maiden Cleuch to House o' Muir on the south-east side—once the scene of famous cattle trysts.

The chief heights in order from north-east to south-west are Allermuir (1617), Black Hill (1628), Carnethy (1890), Scald Law (1898), West Kip (1806), East Cairn (1839), West Cairn (1844), Mount Maw (1753), Craigengar (1700), Byrehope Mount (1752). As they pass to the south the hills decrease in altitude, and they are connected by an irregular group of heights with Broughton Hills in Peeblesshire, which may be regarded as the commencement of the Southern Uplands.

Glencorse Burn or Logan Water, the Water of Leith and its tributaries, and Lyne Water flowing to Tweed are the main channels of Pentland drainage. The streams and springs in many cases have been impounded to supply water for the needs of the metropolis and thus various reservoirs have been formed.

The rounded hillslopes are in some parts bleak and covered with heather : in others they are clad with grasses which afford splendid pasture for sheep.

Several of the gullies which the hill streams have carved in the range are remarkable for their romantic beauty.

On Threipmuir, says tradition, Sir William St Clair of Roslin staked his head on the fleetness of his hounds and, having *threeped* the dogs effectively to a kill, he won a grant of the Forest of Pentland from King Robert the Bruce, a success which led St Clair to found, near by, the Chapel of St Katherine, the ruins of which are now covered by the waters of Glencorse Reservoir.

Threipmuir Reservoir

The Moorfoot Hills occupy a large part of the south-east corner of the county and stretch into the adjacent parts of Peeblesshire. The valley of the Gala separates them from the Lammermoors, the western slopes of which run along the south-east border of the county. Both topographically and geologically the Moorfoots mark the boundary between the Central Valley of Scotland and the Southern Uplands. Resulting from the denudation of a triangular tableland, they form two broken lines of isolated hills and groups of summits, usually rounded and rolling in outline, and of bleak moorland character suited only to the pasturing of sheep. They reach their highest point in Blackhope (or Blakeup) Scar (2136 ft.), the loftiest ground in the county. Gala Water and its affluents, Heriot Water, and Luggate Water, drain their eastern flank into Tweed, while the South Esk and the South and the North Middleton Burn, which unite to form the Gore, itself a tributary of the South Esk, are the main streams on their north-western slope, carrying their waters to the Firth of Forth.

The axis of the Pentland ridge is continued through the Braid Hills and Blackford Hill towards the Forth. "The furzy hills of Braid" reach a height of 698 feet, and from them there is a noble prospect of the rich champaign country stretching away to east and west, and of the southern districts of the city guarded by the couchant lion of Arthur's Seat and rising to the Rock,

> "Where the huge castle holds its state,
> And all the steep slope down,
> Whose ridgy back heaves to the sky,
> Piled deep and massy, close and high,
> Mine own romantic town!"

On the hills of Braid, where Johnnie of Braidislee once hunted the dun deer, the city golfers in their hundreds now enjoy a milder sport.

The Braid Burn has cut a deep and picturesque glen between the Braids and Blackford,

> " on whose uncultured breast,
> Among the broom and thorn and whin,
> A truant boy, I sought the nest,
> Or listed, as I lay at rest,
> While rose, on breezes thin,
> The murmur of the city crowd,
> And, from his steeple jangling loud,
> Saint Giles's mingling din."

Of the numerous hills of volcanic origin with which the plain of the Lothians is dotted, specially interesting are those that ring round the city of Edinburgh. The whole of Arthur's Seat (822 feet) is included in the King's Park, and the Queen's Drive encircles it at varying altitudes. The basaltic columns known as Samson's Ribs form a lofty cliff on the southern side of the hill, and on the west Salisbury Crags sweep round in a bold curve of precipitous rock with a smooth straight slope of grass-covered detritus descending to low ground beneath. From the south-west the hill presents the familiar aspect of the "Lion Couchant."

The Calton Hill (348 feet), with its many public monuments and buildings, is the Mars Hill of the Modern Athens, as the Castle Rock (437 feet) is the Acropolis.

To the south-west rise the twin heights of East and West Craiglockhart (550 feet), named from the old peel tower, built by Lockhart of Lee, the ruins of which still stand near the western base. The summit and slopes of the hill are now occupied by a lunatic asylum, a poor-house, a fever hospital, a hydropathic establishment, and a golf course.

Arthur's Seat

Corstorphine Hill (520 feet), finely wooded and crowned with Clermiston Tower, a memorial of the Scott Centenary celebrated in 1871, is a conspicuous feature on the western outskirts of the city, and "Rest-and-be-thankful" on its eastern slope is a favourite view point.

Ratho parish, "the place of the raths or hill forts," has two abrupt eminences similar in character to the fore-going—Kaimes Hill and Dalmahoy Crag (680 feet), twin heights cut off from the Pentlands by the valley of Leith Water, fronting precipitously towards the west, and form-ing a prominent feature in that part of the Lothian plain.

5. Watershed. Rivers. Lakes.

The general slope of the county, as already described, is to the north and east; and the drainage accordingly is mostly carried to the Forth from the Pentlands and the Moorfoots, which form the main watershed. The south-east corner however, the parishes of Heriot and Stow, are in this respect cut off from the rest of the county, the drainage of the south-east slope of the Moorfoots and the south-west side of the Lammermoors being carried south into Tweeddale. On the northern plain the chief streams, in order from west to east, are the Almond, the Water of Leith, the Braid Burn, the Sten-house Burn, the Esk, and the Tyne: Gala Water gathers the drainage of the southern slope.

The Almond flows through the county for only a few

miles in the middle part of its course.   To begin with, it
is a Lanarkshire river, rising at an altitude of 700 feet in
the parish of Shotts.   Flowing in an easterly direction
through Linlithgowshire, it touches the boundary of Midlothian near the village of Livingston.   After passing
through part of Midcalder parish it again forms the
boundary between Mid and West Lothian till it reaches
the Firth of Forth at Cramond.   The boundary between
these counties in the southern part is formed by Breich
Water, a tributary of the Almond, which joins the latter
at the point where it first touches Midlothian.   Two
good trouting streams are Linhouse Water from the
Pentlands, which joins the Almond on the right near
Midcalder, and the Gogar Burn which, after a winding
course from near the middle of Kirknewton parish, falls
into the Almond near Turnhouse.

The length of the Almond from source to sea is about
24 miles.   The lower part of the course is winding,
through flat and fertile country, which was subject to
flooding until the river banks were raised and strengthened.   Near the sea the Almond cuts through a ridge in
a deep, finely wooded gorge.   Between Midcalder and
Kirkliston the Union Canal crosses the stream in an
aqueduct.   Further down, the valley is spanned by the
lofty viaduct of the Edinburgh and Glasgow branch of
the North British Railway.

The Water of Leith, "the water of the hollow," the
head streams of which rise in the Pentlands near the
West Cairn Hill, 1400 feet above sea-level, flows in a
north-easterly direction for about 23 miles to the sea

at Leith Harbour. Its course is thus fairly rapid, and this
fact has led to its being so much used as to merit the description of the hardest worked river in Scotland, "a most
serviceable drudge that is by no means spared."   In 1793
there were no fewer than 80 mills of different sorts on
ten miles of its course; and though the number is much
decreased, corn and flour mills, snuff mills, and paper mills
still make considerable use of its current.   At one time
also these industries made the stream the vehicle for con-
veying away their refuse, and the city of Edinburgh likewise put it to such base uses.   But the Water of Leith
Purification Scheme, carried through near the end of last
century, has changed all that, and trout may now be angled
for in its waters even where they flow through the city.

The chief tributary is the Bavelaw Burn which joins
on the right bank at Balerno: and there are numerous
smaller streams draining into it from the north slope of
the Pentlands, along whose base Leith Water flows.

In spite of the industrial character of the valley, the
scenery is charmingly varied and picturesque.   The bleak
moors and green hills at its source, the deep rocky channel
it has carved for itself at the foot of the hills, the open
champaign of its lower course, the romantic gorge within
the city bounds—each has its peculiar beauty and each is
rich in historic association.

The Braid Burn rises in the Pentlands, one and a
half miles from Bonaly Reservoir, at a height of about
1000 feet above sea-level, and after a course of nine
miles in a north-easterly direction reaches the Firth of
Forth as the Figgate, so named from the Figgate Whins

Hawthornden

which at one time covered its banks in the lower part of its
course. It passes through two picturesque gorges, one at
Dreghorn, and the other between the Braids and Blackford
Hill, the latter containing the Hermitage of Braid, lately
the home of Sir John Skelton, the well-known "Shirley,"
able defender of Mary Stuart.   Duddingston Mill is the
only one now remaining of three whose owners in 1789
opposed the project of the Town Council of Edinburgh
to increase their water supply at the expense of the Braid
Burn.

Burdiehouse or Niddrie Burn has its source on the
northern slope of the Pentlands near Dreghorn, flows
east and north-east for eight and a half miles and
empties into the Firth of Forth at Magdalen Bridge,
near Fisherrow.

The Esk is formed by the junction near Dalkeith of
the North and South Esks.   The North Esk rises near the
Boarstane on the northern slope of the East Cairn Hill in
the parish of Linton, Peeblesshire.   For two and a half
miles it forms the boundary between that county and
Midlothian, and then turning to the north-east it flows
17 miles to its junction with the sister stream.   Habbie's
Howe, charmingly described in Allan Ramsay's pastoral
drama *The Gentle Shepherd*, is identified with the valley
of the Esk near Carlops; but the most famous part of its
course is the romantic glen of Roslin and Hawthornden,
noted alike for its picturesque scenery and for its historical
and literary associations.   The finely wooded grounds of
Melville Castle, below Lasswade, and the noble park of
Dalkeith continue the pleasing features of its course.

River Esk, Roslin Glen

The source of the South Esk is at a height of 1700 feet on the western side of Blackhope Scar, the highest of the Moorfoot Hills. The river flows for 19 miles in a northerly direction to the meeting of the Esks. Its chief affluents are Fullarton or Redside Burn, Gore Water, and Dalhousie Burn. The scenery near Dalhousie Castle and in the park of Newbattle Abbey with its great beeches, rivals that of the north branch; while the basin contains rich coalfields, giving employment to the miners of Gorebridge and Newtongrange.

The conjoined river continues its course through the park of Dalkeith Palace, breaks through the ridge on which stands the church of Inveresk, and flows into the sea between Fisherrow and Musselburgh. The old bridge at Musselburgh, if not Roman, is at all events of considerable antiquity.

The Tyne, which has a course of five miles near the middle of the eastern boundary of the county, belongs to East Lothian. Gala Water, rising among the Moorfoots and following a winding course to the south-east, carries the drainage of Heriot and Stow parishes to the Tweed. The chief tributaries of the Gala are Heriot Water and Luggate.

Most of the lakes in the Pentlands and the Moorfoots, as already noted, are artificial and have been constructed with the object either of providing the city with water, or of compensating proprietors with rights of use over waters affected by such interference with the natural drainage.

For the former purpose Glencorse Reservoir was

formed in 1819–1828. It is crescent-shaped and has much of the beauty of a natural loch. Threipmuir and Harelaw were constructed in 1847–1848 to maintain supply for millowners on Bavelaw Burn and Water of Leith. Between 1850 and 1868 Clubbiedean, Torduff, Loganlee, Bonaly, Harperrig, and Crosswood were added to the list. In 1869 the springs of the Moorfoots were drawn upon; and Gladhouse, Rosebery, and Edgelaw Reservoirs, all in the basin of the South Esk, were formed. Near the south-west boundary with Lanarkshire, Cobbinshaw Loch was made to supply water to the Union Canal.

In the immediate vicinity of Edinburgh are several natural sheets of water, the dwindling remains of the more spacious lakes of prehistoric times. In the King's Park are Duddingston, Dunsappie, and St Margaret's Lochs; and near Restalrig is Lochend. Duddingston Loch lies at the south-eastern base of Arthur's Seat, and, with its swans and other wild fowl, with its old boathouse on the promontory crowned by the ancient church and tower, and with the bold red crags for background, it makes a fine picture on a summer evening; but it is even more striking in winter when covered with ice and crowded with skaters and curlers from the city. From its bed many relics of antiquity have been recovered.

Dunsappie curves round the foot of a crag on the east shoulder of the hill near the line of the upper Queen's Drive.

St Margaret's is a small sheet of water at the bottom of the hill on the north side.

Duddingston Loch

Lochend is situated near the village of Restalrig, and on the cliff above it once stood the castle of Logan of Restalrig. The loch was formerly much more extensive; and in 1871 vestiges of the framework of an ancient lake dwelling were discovered near it. It was at one time the source of water supply for Leith.

Besides those mentioned, several other lakes are known to have existed in the neighbourhood of the city; and of these, two or three persisted till comparatively recent times. A great lake extended from Gogar to Corstorphine and was continued to the outskirts of the town at Haymarket. There it may have communicated with the Burgh Loch, which covered the site of the Meadows, and with the Nor' Loch, which encircled the northern base of the Castle Rock. The bed of the Nor' Loch is now occupied by the North British Railway and by Princes Street Gardens.

6. Geology and Soil.

The crust of the earth is composed of *rocks*—which name is applied by geologists not only to the hard coherent material popularly so called, but also to soft and loose substances such as clay or sand. All these substances have resulted from the cooling of the gaseous mass which was the original form of our planet. Obviously the cooling would begin on the outer surface of the glowing globe; and in the second stage there would be a cold rind of solid matter enclosing a core at a much higher temperature.

The cooling was necessarily accompanied by contraction, and the contraction resulted in a crumpling and folding of the crust comparable with the wrinkling in the peel of a dry apple. Moreover, the interior of the earth was subjected to such enormous pressure through this contraction of the crust that, although its temperature must still be a white heat, its mass is calculated to be more rigid than solid steel: and if any weakness occurs in the enclosing crust the internal content is apt to be forced out to the surface.

Surrounding the solid earth there was formed the atmosphere, and this contained much water vapour which condensed on the surface of the Earth and filled its hollows as the Ocean.

On account of the crust movements caused by contraction and internal pressure, the water has not always occupied the same positions on the surface. At one time huge tracts have been heaved up: at another, they have sunk, the seas have flowed over them, and the solid matter of which these tracts are composed has been sifted and re-arranged largely through the agency of the water.

Rocks are accordingly classified under two great divisions, (1) igneous or primary—those that have resulted directly from the cooling and solidifying of the gaseous mass; (2) aqueous or sedimentary or secondary—those which have been formed by the destruction of the primary rocks and the re-deposition of their materials chiefly through the action of water.

Of the igneous rocks some have cooled on the surface; others have cooled at considerable depths beneath the surface; and the deeper the cooling, the greater has been the pressure under which the rock has formed and the harder and more crystalline is its structure. Thus lava is an igneous rock formed at the surface, while granite is also igneous but has solidified at a great depth and is highly crystalline in structure. The effect of pressure is also frequently shown by *folding* as in the case of mica schists and Silurian rocks.

Sedimentary rocks are characterised by *bedding*, i.e. they readily split into beds or layers. This character is due to the manner of their formation, layer after layer having been deposited on the bottom of the sea or lake, just as one might lay sheet upon sheet of paper. In most instances, accordingly, we can tell by the relative position of the layers which stratum of aqueous rock was first laid down and is therefore oldest in time : since if the original position has remained unchanged that which is nearest the surface must be the youngest. In a similar way geologists are able to make a rough calculation of the age of rocks ; for the same process can be seen going on at the present time and the rate of deposition estimated. Moreover, these secondary rocks often contain fossil remains of animals and plants which were living at the time when the rocks were being formed, and these fossils give indications of the conditions, climatic and other, then prevailing. Indeed palaeontologists—those who have studied these remains of ancient life—have classified secondary rocks by the fossils found in them. The oldest of all, i.e. the lowest, is named Primary or Palaeozoic ; then come in ascending order Secondary or

Mesozoic; Tertiary or Cainozoic; and lastly, Quaternary or Post-Tertiary.

It is not to be supposed that all of these systems are represented in all parts of the earth's surface; and a marked peculiarity of Midlothian is the absence of both the Secondary and the Tertiary group. There is nothing between the rocks representing the Primary group and the most recent formation—the Post-Tertiary or Quaternary.

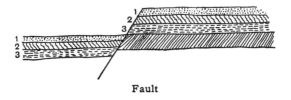

Fault

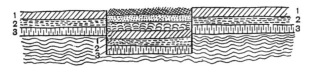

Formation of Rift valley

Two results of the crust movements already referred to are of special interest in Midlothian. Rock layers are sometimes split through by a crack extending for a long distance in the strata; and a tract of the surface may slip down or be heaved up relatively to the adjacent tract. Then a *fault* is said to occur. In the figure above layers 1, 2, and 3 ought to have been continuous, but a slip has occurred and non-conformity has resulted.

A slip may be due to parallel faults and a wide portion of surface be affected while adjacent areas on either side may remain at the original level. In this way a rift valley is formed as in the figure on p. 25.

This is what has happened in Central Scotland. A great rift valley extends in a south-westerly direction from Kincardineshire and Haddingtonshire on the east to Renfrewshire and Ayrshire on the west. The Central Valley owes much of its industrial prosperity to this subsidence : for it has preserved in its depths the coal-bearing strata which once extended on either side and which have been entirely denuded from these adjacent areas.

Another effect of the pressure consequent upon the contraction of the surface is the folding or rippling of the strata, just as a sheet of paper is rippled by pressing the edges in towards the middle. The upfolds so formed are known as anticlines and the downfolds or troughs as synclines ; and both of these formations are illustrated in the strata of the county. Thus the Pentland Hills form an anticline of Upper Silurian Rocks (shales and grits), over which the Carboniferous Rocks were upfolded. Most of the latter have been denuded, but coarse conglomerates, grits, and sandstones with sheets of felstone and ash belonging to the Old Red Sandstone group still remain. The continuity of the upfolding strata is also broken by faults parallel with the axis of the Pentland uplift.

On the east of the Hills extends the Midlothian coalfield with its Carboniferous strata rising on the west against the Silurian rocks of the Pentlands, on the south against the Silurians of the Moorfoots, and on the east against the Carboniferous Limestone of Roman Camp Hill, which itself is an upfold sinking again into the Coal Basin of East Lothian. On the outer edges of the basin the strata are inclined at such a steep angle that the coal seams, which are necessarily the lowest and oldest, are spoken of as "Edge" Coals, as at Edmonstone. In the middle of the basin the seams are flat.

To the west of the Pentland uplift the surface consists also of Carboniferous Rocks arranged in a series of folds and broken into every here and there by isolated masses of intrusive igneous rocks, chiefly greenstone and basalt.

In the Midlothian Coalfield, if we suppose ourselves to strip off layer after layer of the strata from the surface inwards, we should find the order to be as follows :

1. Soil, Sand, Glacial Clay, etc.
2. Upper Coal Measures.
3. Millstone Grit.
4. Carboniferous Limestone series.
5. Lower Carboniferous group.
6. Upper Old Red Sandstone.
7. Upper Silurian.
8. Lower Silurian.

As has already been noted, with the exception of No. 1 of the above, which belongs to the Post-Tertiary period, these strata are included in the Primary or Palaeozoic group.

The superficial drift, clay, sand, and gravel, consists in large part of boulder-clay or till—a stiff sandy clay, with stones, more or less rounded, embedded in it. This was deposited, during a period of submergence, by the action of drift ice. It is found all over the county and reaches a depth of 50 to 100 feet.

The Coal Measures are in two series, interstratified with sandstone, fireclay, clay ironstone, and shale. They extend to a depth of 1220 feet, and rest upon the Millstone Grit, a coarse red or white sandstone 340 feet thick. This is succeeded by the Carboniferous Limestone series, consisting of beds of limestone, associated with coal, ironstone, fireclay, sandstone, and shale.

The Lower Carboniferous series was laid down during a period of volcanic activity, as is shown by the contemporaneous occurrence of numerous extruded Trap Rocks.

The Upper Old Red Sandstone is found along the central axis of the Pentlands. It contains the igneous rocks, formed at the same time, which constitute the prominent features of the Pentland Chain.

The Upper Silurian Rocks occur in patches entirely within the range of the Pentlands. They are, so to say, the foundation rocks of the chain, and are covered unconformably by the previously named series.

The Lower Silurian Rocks are part of the edge of the Silurian region which forms the tableland of Southern Scotland. One portion of them is found overlapped by the coal measures near Penicuik, and another forms part of the Moorfoot Hills.

While these groups or series of rock beds may be regarded as the normal stratification of the rocks in the county, more striking from the physical point of view are the intrusive igneous rocks which have been thrust through and between them. These are known to be contemporaneous with the Silurian series, as in the case of a patch of greenstone at Bavelaw; with the Old Red Sandstone, as in the felstones of the Pentlands; and with the Carboniferous series, as in the trap of the Castle Rock, St Leonard's Hill, Salisbury Crags, the Dasses, Corstorphine Hill, Ratho Hill, Dalmahoy, and the Kaimes.

Crag and Tail

It is probable that there were active sub-aerial volcanoes at Arthur's Seat, the Braid Hills, and in the northern Pentlands. In most cases the igneous material was thrust, not through, but into the superincumbent strata, and subsequent denudation has removed the softer aqueous beds and has left the harder volcanic rock exposed, as in the case of the Castle Rock.

This rock is a fine specimen of a feature very prominent in the district—Crag and Tail. The crag rises steeply from the low ground which extends to the

west, while on the eastern side there is a gentle slope
from the summit of the crag to the level of the surrounding
plain.   Evidently this has resulted from the wearing away
of the softer material by some agency moving in an
easterly direction.   The resistant volcanic rock withstood
this wearing agency and protected the softer rock to the
east, which forms the "tail," now covered by the houses
of the old town.   Similar formations occur in the Calton
Hill, which consists of volcanic ash supported by beds of
lava; in Corstorphine Hill, where basalt stands out from
the sandstones surrounding it; in Blackford Hill, whose
porphyritic rock presents a steep face to the west and
south, backed on the eastern side by a long slope of the
protected stratified rocks.

The soil of the county varies considerably in fertility.
There is a marked absence of the calcareous soils which
are so important to the agriculturist of the north of
England.   Some of the hills are moorish and mossy;
others are covered with a thin clay.   In the valleys and
river basins a deep and rich loam prevails, which has been
greatly improved by cultivation and draining.   Near
Corstorphine a black loam of great fertility makes that
part of the county the garden of Edinburgh.   The
northern and central portions of the county are the most
fertile; the south and south-east are mainly pastoral;
and about one-third of the area is not arable.

7. Natural History.

Shallow waters surround the British Isles. Nearly all the North Sea is under 100 fathoms deep, and therefore if the sea bottom were raised 600 feet, less than the height of Arthur's Seat, it would become dry land, and these islands would be joined to the continent of Europe. This would merely be repeating what geologists assure us was the case in ages long gone by; and their opinion is corroborated by the botanist and the zoologist, who point out that the various species of plants and animals existing in Britain and Ireland are practically identical with those on the mainland of Europe. As compared with the latter, however, the British Isles are deficient, especially in mammals, reptiles, and amphibians; the reason being that the final separation took place before many of these land animals had extended their range to Britain. On the other hand the isolation has resulted in the occurrence of a few species which are peculiar to the islands, as the red grouse, and certain fishes such as the vendace of Lochmaben.

Change in climatic conditions is naturally accompanied by change in animal and vegetable life. At one time the climate of the region was similar to that of the sub-arctic countries of to-day; and the plants and animals were then such as are now found in Northern Europe, Asia, and America. In Duddingston Loch and also at Cramond remains of moose and elk have been found embedded in mud; bones of the mammoth or hairy elephant have been unearthed at Clifton Hall, near Ratho; and some years ago

those of reindeer and wolf were dug up from a cleft in
the rocks on the Pentland Hills near Dreghorn. Evidence
too has been noted in the Edinburgh area of the existence
at a later date of the brown bear and the wild boar. But
human settlement gradually encroached on the domain of
the wild animals, and the fiercer and more dangerous were
hunted down and exterminated.

The smoke and dust of industrial centres have also
had their effect in limiting the range of species. Thus
certain butterflies formerly known in the area are no
longer observed; the famous *Hesperia Artaxerxes* has not
since 1868 been captured on Arthur's Seat, which once it
haunted.

Many of the ponds and lochs in the county are peculiarly rich in the lower forms of life. The Upper Elf
Loch on the Braid Hills is noted for the number and
variety of its micro-fauna; 69 protozoa, two coelenterata,
and 97 rotifera having been identified as existing in its
waters. In the canal at Slateford certain fresh water
sponges are to be found.

Omitting mention of intermediate orders such as those
of spiders and insects—more than 3800 species of the
latter have been observed in the area—we may note that
trout are numerous in the streams of the county, and
that the Water of Leith Purification Scheme has once
more made angling possible even within the bounds of
the city. Salmon are occasionally caught in the Esk.
Pike, perch, and eels live in Duddingston and other lochs.
Loach are common locally and are found in Braid Burn,
Lothian Burn, and Threipmuir Reservoir.

Of reptiles three species are known to live in the district; the viviparous lizard, which is to be met with on Bavelaw Moss; the blind worm or slow worm, which was lately to be found on Blackford Hill and has even been seen in Princes Street Gardens; and the adder, which has been occasionally taken on the Pentlands.

Six species of amphibians, the common frog, two toads, and three newts, have been noted.

The bird life of the district is rich and varied, 248 species of land, shore, and sea birds having been observed. This variety is mainly due to the position of the area, fronting the North Sea and in a line of migration. Of the 125 species of birds which breed in the area, 68 are residents and 31 are summer visitants. Of the 123 non-breeding, 58 are winter visitants, 28 are birds of passage, and 37 are casuals. The osprey, the hen harrier, and the black guillemot have disappeared in recent years; but, on the other hand, the great crested grebe, the great spotted woodpecker, the hawfinch, the pied flycatcher, and certain ducks have been added to the list of breeders. The beautiful plumage of the kingfisher is occasionally to be seen glancing above the stream in the Water of Leith valley. Several kinds of owl are observed; larks, thrushes, and fieldfares are numerous, and starlings are increasing in number. The sparrowhawk and the kestrel are frequently noticed. In autumn and winter the golden plover is a regular visitor, while the lapwing or peewit lives in large numbers in the district all the year round. Gulls come far inland in search of food. Pheasants, partridges, and woodpigeons are common; the stock dove is widely distributed; and quail are occasionally observed. Swifts, swallows, sand martins and house martins are still abundant in summer; the various tits are common; finches and linnets are plentiful; and sparrows are so numerous as to call for special effort in keeping them in check as farm pests.

For the whole of Scotland 57 species of mammals are recorded; and of these 50 have been found in the Edinburgh area. These include three species of bat; three shrews, the hedgehog, and the mole; wild-cat, pole-cat, weasel, and stoat or ermine; squirrel, water-vole, field-vole, bank-vole, brown and black rats, house, field, and harvest mouse, hare, mountain hare, and rabbit; and red, fallow, and roe deer. The wild-cat, the pole-cat, and the marten are now extinct; but the badger has been re-introduced and known to breed in Midlothian in recent years; while otters not so long since used to haunt the Braid Burn near Duddingston and the Burdiehouse Burn at Niddrie.

In all, close on 7000 species have been recorded as included in the fauna of the Forth area.

Presenting as it does a considerable variety of surface differing in geological structure and soil, elevation, and exposure, the county affords to the botanist a not uninteresting field. The shores of the Firth of Forth, the valleys of its tributary streams, the cultivated plains, the numerous woodlands, the wide moors, and the hills have each their characteristic plant associations, and the flora of the district is accordingly of wide range and great variety. More than 400 genera, over 1000 species and varieties of flowering plants have been enumerated as occurring in the locality; the list of ferns and their allies contains 18 genera and 43 species and varieties; while 520 species and varieties of mosses, liverworts, lichens, and charas have been noted.

The short coast-line is so much occupied by towns, and harbour and other works, that little space is left for the growth of plants characteristic of such a locality: but various species of algae and of plants inhabiting salt marshes and sandy shores are to be found here and there on the coast.

Duddingston Loch, which is the largest natural lake in the county, abounds in the common reed (*Arundo Phragmites*) which is a typical marginal plant forming a swamp. In or near the loch are also found the celery-leaved buttercup, the great spearwort, mare's tail, yellow flag, marsh- and water-speedwell.

On the cultivated lowlands many of the species found have been introduced in recent times in association with various crops.

Of the trees the chief indigenous species are the oak, the birch, the alder, the wych elm, the ash, and the Scots pine: but most of the woods now existing have been planted by man, and many foreign trees have been introduced; such as beech, chestnut, lime, sycamore, larch, and spruce. The upper woods in the Esk valley, for example, were all planted on moorland above 150 years ago; below Penicuik they are comparatively recent. Small natural woods occur, but these too are generally altered by the introduction of planted trees. In the Esk valley grow practically all the native and introduced trees to be found in Scotland.

Deciduous trees occupy the lower altitudes; the coniferous woods cover the hill slopes at greater elevations. Of the former type, besides those previously named, holly, sloe, gean, rowan, hawthorn, elder, hazel, willows, and poplars are all more or less common. The Scots pine is the commonest of the conifers, but larch and spruce are also grown.

The valley woodlands are associated with special types of plants, among which may be named as interesting, because uncommon in the area, the bird's nest orchid, toothwort, giant horsetail, and cuckoo pint.

On the basaltic hills (Arthur's Seat, Blackford, the Braids, and Dalmahoy) certain plants seem to find a peculiarly suitable habitat. Such are maiden pink, red catchfly, spring sandwort, bloody cranesbill, spindle tree, and forked spleenwort (*Asplenium septentrionale*).

In the locality of the conifers, heather begins to appear, and large tracts of the hill country are covered by it or by various types of grassy heath or moor. Heather associations may be classed into heaths, heather moors, and sphagnum moors according to the amount of moisture present in the soil, the sphagnum moor being the wettest and most peaty. Examples of all three are to be found in the Pentland area.

A few Alpine plants have lingered on in certain places on the Pentlands, relics of a time when the climatic conditions were much more severe. Of these may be mentioned *Schollera Oxycoccus*, *Juniperus communis* (rare in

Midlothian) and *Rubus Chamaemorus* or cloudberry, which last is generally regarded as typical of arctic-alpine vegetation on peat.

8. Peregrination of the Coast.

The coast-line extends for about twelve miles, from the mouth of the River Almond in the west to the boundary with Haddingtonshire, a little beyond Levenhall, on the east. The shore for the most part is either sand or shingle, low, and shelving gently seaward. Here and there, as at Wardie Bush, off Leith Harbour, at Joppa, and at West Pans, occur stretches of low waterworn rocks.

From 25 to 30 feet above the present mean tide mark, a raised beach skirts the shore line throughout its entire length; and on it are built Musselburgh, Fisherrow, part of Portobello, and part of Leith. It includes also Leith Links, famous at one time as a resort for Edinburgh golfers, royal and other. Further inland and parallel with the coast, runs a ridge which appears to represent an ancient sea-cliff. It begins on the east beyond the confines of the county; and, at a height of about 100 feet above present sea-level, it curves behind Musselburgh and Portobello, passes through Edinburgh, and continues to the north-west behind Granton and Muirhouse. This ridge has been important historically; on it in East Lothian was fought the Battle of Prestonpans; in Midlothian, the Battle of Pinkie.

Off Cramond, sandy flats, with beds of mussels, cockles, and other shellfish, extend beyond Cramond Island so that it is possible at low water to walk across to the island, which lies three-quarters of a mile off shore. The island is large enough to afford room for a farm, on which are grown excellent potatoes. Inch Mickery is a still smaller islet, one and a quarter miles to the north.

Cramond itself, "the fort on the river," a picturesque little village, occupies the site of a Roman station of some importance, evidence of this being afforded by the coins, medals, and other Roman relics which have been discovered in the neighbourhood. At one time it boasted a harbour accessible to small vessels, but that has long been silted up; one or two little pleasure yachts and a few rowing boats are now all its fleet. A free ferry over the Almond is maintained by Lord Rosebery and is much used by pedestrians taking the private path through Dalmeny woods to Queensferry. In the grounds of Cramond House an old tower still remains, all that is left of the palace of the Bishopric of Dunkeld, which was once occupied by the poet-bishop, Gavin Douglas.

About a mile inland on the rising ground behind the old beach mentioned above, is situated Lauriston Castle, once the property of the celebrated financier John Law (1671–1729), on whom was written the witty epitaph—

> "Ci git cet Ecossois celebre,
> Ce calculateur sans egale,
> Qui par les regles de l'algebre
> A mis France à l'hôpital."

Cramond Brig

West of Granton Harbour the policies surrounding the "Tudor" mansion of Muirhouse come down to the shore in a well-wooded slope. Two towers of an older house built about 1670 still remain.

Granton Quarry, from which was excavated the stone used in the construction of the harbour and breakwaters, after being breached by the sea was used for some years as a station for investigating marine life.

Close to Granton is the fine old house—Caroline Park, built in 1685 by George Mackenzie, Viscount Tarbat, when he was at the head of affairs in Scotland. It was afterwards the property of John, Duke of Argyll, who named it in honour of Caroline, the sagacious Queen of George II; but it has now fallen on evil days and is put to prosaic use as a factory of ink.

From Granton to Trinity the shore is protected by a strong sea-wall. Trinity consists to a considerable extent of villa residences and had at one time—till it was destroyed by a storm—a chain pier much used by Edinburgh bathers. The town is now continuous with Newhaven, "Our Lady's Port of Grace," which was famous as a harbour and shipbuilding yard in the reign of James IV, when "ane varie monstrous great schip," the *Michael*, the largest in Europe at that time, was built, using up so much timber that "she waisted all the woodis in Fyfe except Falkland wode." The harbour is now entirely given up to the fishing industry, and the inhabitants of the town are a picturesque and peculiar people, the fishwives in their quaint and characteristic dress being a familiar feature in the streets of the capital.

Granton, Newhaven, and Leith are now practically one long town skirting the shore and linked by a line of electric tramways. Off Leith east pier, lights mark the Black Rocks, but the beach continues low and sandy past Seafield and the Craigentinny Meadows—at one time irrigated and enriched by Edinburgh sewage and now

Portobello Promenade

partly occupied at their eastern end by the Marine Gardens, a popular summer resort on the outskirts of Portobello. Near this a prominent monument stands close to the Edinburgh road. It was erected by a former owner of Craigentinny estate and has panels of finely sculptured marble with representations of the Israelites crossing the Red Sea and Miriam's song of triumph.

A fine stretch of sands, an extensive promenade, and a pier make Portobello a popular summer resort. The site of the town was formerly a whin-covered waste known as the Figgate Whins. This wilderness is said to have afforded a hiding-place to William Wallace when he was on his way to attack Berwick, and at a later time was much resorted to by less reputable characters, smugglers and footpads. The first house was built by a seaman who had served on the Spanish Main and who named his home after the Spanish town captured by Admiral Vernon in 1739.

On Portobello Sands a fruitless conference took place between Cromwell and the Scots leaders during the campaign which culminated at Dunbar Drove; there, too, Bonnie Prince Charlie arrayed his forces before the march to Derby; and the Sands were also the scene of some of Sir Walter Scott's military experiences as Quartermaster of the Edinburgh Light Horse.

At Joppa occurs the group of low rocks already noticed, and immediately to the east is the hamlet of Joppa Pans, so named from the salt works, once an important industry there and at Pinkie Pans, and still represented. "Salters," as the workers were called, were classed with colliers as being *adscripti glebae*, bound to the estate as serfs, which condition was maintained in many cases even up till the beginning of last century.

At Magdalen Bridge, where an old toll called the "Gentes Custom" used to be levied on beasts of burden, the Burdiehouse Burn passes under the coast-road and reaches the sea.

The harbour of Fisherrow is situated about half a mile west of the mouth of the River Esk, at or near which point there appears to have been a port of a kind ever since Roman times. Only boats of light draft can make use of the harbour even at high water. On either side of Eskmouth there are fine stretches of level links, those on the east or Musselburgh side being famed for archery, golf, and horse-racing. For two months in 1650 Cromwell's infantry encamped on the links and his own tent stood near Linkfield.

At this point the sea is going back and in recent years means have been taken to extend the links seaward. The mussel bank from which the town takes its name lies off the river mouth. *Brogh* or *burgh* in this case is said to mean mussel bed, and hence the significance of the old rhyme—

> " Musselburgh was a burgh
>   When Edinburgh was nane;
>   An' Musselburgh 'll be a burgh
>   When Edinburgh has gane."

At Levenhall there is again an outcrop of rock on the beach ; and the Ravenshaugh Burn, the boundary between Mid and East Lothian, has cut a deep dell in the cliff which here overhangs the road leading east to Prestonpans and Dunbar.

9. Climate and Rainfall.

The main factors influencing climate are latitude or distance from the equator, height above sea-level, and position as regards the sea. In the case of the British Isles all these factors tend to produce a temperate climate. Our islands lie in that belt of latitude which is known as the Temperate Zone: their general land-level is of moderate altitude: and the ocean circles them round so that no spot within them is at any great distance from it.

But apart from these general considerations, certain special conditions still further modify for us extremes of temperature. Within the latitude 50° to 60°, in which our islands are placed, there is a fairly constant drift in air and ocean from the south-west, resulting mainly from two causes—difference of temperature in different tracts on the earth's surface, and the west-to-east rotation of the earth. Warmth and moisture are carried by winds and ocean currents; and, inasmuch as the south-westerly drift comes from the warmer regions nearer the equator, we find that the western sides of continents are both warmer and wetter than the eastern sides: or, to put the same statement in another way, the eastern shores of the great oceans are warmer than the western. Thus while the coast of Labrador is ice-bound in winter, our harbours—in the same latitude—are open all the year round.

While, as has been hinted, this south-westerly drift is fairly constant, it is not to be supposed that in such an unstable body as the atmosphere, changes and variations are not constantly occurring. Winds are the result of differences of atmospheric pressure. Just as gravitation causes water to seek a lower level, so the same force causes air to flow from a region where pressure is high to one where it is comparatively low. The atmosphere is thus constantly moving in huge waves, the crest of a wave causing a high pressure, while the trough is an air region of low pressure. These wind movements are known as anti-cyclone and cyclone respectively; the former, indicated by a high barometer, is accompanied by light airs, clear sky, and fine weather; the latter sends down the barometer and brings strong winds, cloud, and rain.

It happens that our islands lie in a regular path of cyclonic disturbances. In the North Atlantic two fairly well-defined areas of pressure are commonly to be found— an area of low pressure in the neighbourhood of Iceland and one of high pressure in the latitude of the Azores. Over the land surface of Europe there is in summer a low pressure centre, and in winter a high pressure area. Accordingly in summer the air is drawn in to the Continent from the Atlantic as a westerly wind, whereas in winter it is drawn more to the north and is therefore a south-westerly wind.

Cyclones have a double or combined motion. In the first place, winds blow in from all sides towards a centre of low pressure and the result is a huge circular eddy which, in the Northern Hemisphere, moves in a direction opposite to that of the hands of a clock. And secondly, the cyclone passes onwards round the world in an easterly course ; so that the combined movement is spiral. Most frequently, we find cyclones coming up out of the Atlantic, striking on the south-west coast of Ireland, and passing over our islands in a north-easterly direction. They have a very considerable influence on our climate ; and as the relations of low pressure to high pressure areas are being more fully observed and understood, the meteorologist is becoming more able to make an accurate forecast of our weather.

Records of wind direction are available for Edinburgh over more than a century, and from these we learn that the mean annual percentage of frequency of winds is as follows : N. 4; N.E. 7; E. 16; S.E. 7; S. 6; S.W. 17; W. 32; N.W. 7; calms and variable breezes, 4. If therefore we take together S.W., W. and N.W. we shall find that winds from these points make up 56 per cent. of the whole : that is to say, on more than 200 days in the year the wind comes from a westerly direction. That is why trees growing in exposed places in our district so often look lopsided, with growth more evident on their east side.

The accompanying figure is intended to exhibit in a graphic way the relative frequency of the different winds. On a compass card showing eight points, circles are drawn at equal distances from the common centre, each circle indicating an additional 5 per cent. All winds from W.N.W. to W.S.W. are called west : all from W.N.W. to N.N.W. are called N.W. and so on. If on the chart we fill in the winds according to the frequency figures already given, we shall have a pictorial representation of the proportional time during which each wind has blown indicated by the size of the segment representing it.

As has been noted, winds from the south-west are

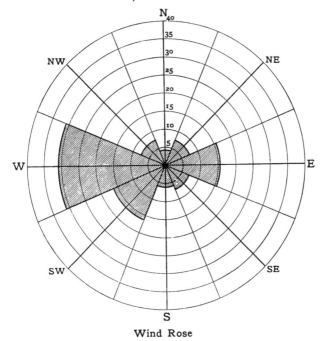

Wind Rose

(showing wind-frequency at Edinburgh)

both warm and moist. We should therefore expect to find that the west side of the country—especially where hills meet the moisture-laden air-current and cause precipitation by forcing it upwards to colder strata of the atmosphere—would have a greater rainfall and be relatively warmer than the east. Accordingly we learn that in the year 1909 the rainfall at Leith was 30·6 inches, at Greenock 60·8, and at Fortwilliam 66·9; while at Kinlochquoich on Loch Hourn, Invernessshire, the wettest station in Scotland for the year, the fall was 112·5 inches. The year 1909 was wetter than usual in Edinburgh district, and the contrast is even more marked in a normal year; the mean annual rainfall at Edinburgh over a long series of years being only 25·42 inches.

It is to be further observed that even within the comparatively small area of the county variation is found. Thus the rule that more rain falls the greater the altitude is exemplified by a comparison of the records for Leith, practically at sea-level, Cockburn Hill near Balerno at 767 feet, and Bowbate in the Moorfoots at an elevation of 2042 feet; the figures being respectively 30·6 inches, 43·5, and 52·09.

For the same year at 57 stations over Scotland the average precipitation was 45·01 inches: for 28 observation stations in Midlothian it was 35·33—10 inches below the general average.

The number of days in the year on which rain falls averages 161: snow falls on 21 days; and hail on 10. In the year there are on the average 6 thunderstorms, and 29 gales.

It is somewhat surprising to be told that the rainfall is least in spring and winter, and greatest in summer and autumn; August being the wettest month of the year and July coming second, while April is driest.

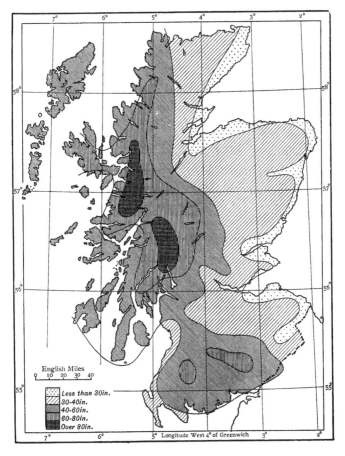

Rainfall map of Scotland. (After Dr H. R. Mill)

Edinburgh is comparatively well off in respect of
sunshine, having a percentage of 31 of the total
possible. The average amount recorded for a year is
1388·1 hours as compared with 1095 at Glasgow and
1401 at Aberdeen. The mean number of days in a
month on which the sun shines for an appreciable time is
23·9. On 15 days in the year fog or mist is experienced;
and the "haar" which is sometimes rolled in by an east
wind from the Firth of Forth is a well-known if somewhat unwelcome visitor to places near the coast. This
same east wind is indeed somewhat notorious in the
district. In spring or early summer it is not at all
uncommon for Edinburgh to have a spell of two or three
weeks east or north-east wind—cold, dry, and bitter;
trying to the stranger within the gates. This phenomenon
is probably due to an extension of the continental high-
pressure area. The east wind is sometimes accompanied
by persistent rain: but it is not unlikely that the rain
may result from the condensation of moisture in an upper
current coming in an opposite direction when it meets the
lower cold easterly current.

In spite of this liability to visitation by the east wind,
Edinburgh enjoys on the whole a pleasant and equable
climate. From observations recorded over fifty years it is
found that the average difference between the mean
temperature of the three coldest months of the year and
that of the three warmest is under 19° F. In other
words the seasonal range of temperature as between
winter and summer is no more than 19°; whereas at
Moscow in the same latitude the range is 49° and in the

interior of Siberia a range of 105° is not unknown. We may take it that this absence of extremes in our local temperature is largely the result of the south-westerly drift of air and sea and of the constantly occurring cyclones bringing to these shores the warmth and moisture of the Atlantic Ocean. Only when the set of the air-currents is reversed are we likely to experience the continental cold or drought. When this happens in winter and we have the anti-cyclonic area of high-pressure spreading outwards to our islands, clear frosty weather supervenes: but whenever the high-pressure wave passes inwards to the mainland again, up comes the warm ocean wind and banishes the frost.

10. People—Races, Population.

Probably the earliest inhabitants of Scotland were short, dark-haired, and long-skulled: they have been called Silures or Ivernians. They were hunters, and used stone weapons, relics of which have been found in various places. Their dwellings were sometimes underground, sometimes in peculiar tower-like structures called brochs, and sometimes on artificial islands or crannogs in the middle of lakes. It is not unlikely that the folk-tales of the mysterious "little people" were founded on memories of this early race. They were conquered by a branch of the great Celtic family—the Goidelic or Gaelic ; but the conquest was rather the introduction of a ruling sept than a complete extinction of the conquered.

The Celtic language became the speech of the people but their physical characteristics remained the same to a great extent. Although it is known that the Celts were a broad-skulled race, the inhabitants of Scotland continued to be of the long-skulled type.

The Goidels were followed by another branch of the Celtic race, namely the Brythons, who occupied the south of Scotland.

Before the coming of the Romans, the Celts of Scotland, to give the inhabitants the name by which they have come to be known, had reached a fairly advanced stage of civilisation, with a knowledge of metals, so that their weapons were of bronze or even of iron.

It is impossible to state exactly the boundaries of the tribes in the south-east of Scotland in the first century of our era. The seaboard from the Tyne (in Northumberland) to the Firth of Forth seems to have been occupied by a division of the Brigantes, known to the Romans as the Ottadeni (or, as in Ptolemy's map, Otalini). To the west lay the land of the Gadeni. From the time of Agricola this region was held by the Romans. Whenever their power was weak, the Caledonians from the north would rush in; and after the middle of the fourth century Roman Britain suffered more and more from assaults by Picts from the north, Scots from the west, and Angles and Saxons from the east. In a short time a wedge of Picts was driven in along the Pentlands; and Teutons from over the North Sea settled along the coast. By the beginning of the seventh century when the various tribes of North Britain were grouped into the kingdoms of Dalriada, Pictland, Strathclyde, and Bernicia, the Lothians were inhabited by a mixed race of Britons, Picts, and Angles.

One result of the mixture of races inhabiting the Lothians is seen in the place-names, in which may be traced the influence of Celts and Angles, and possibly even, to a slight extent, of the earlier race. Thus Pentland is more properly Pechtland, the land of the Pechts or Picts; and Dalmeny may be Dun Mannan, the fort of the Picts of Mannan, who left their name also at Slamannan and Clackmannan. There is a Chesterhall in Midlothian, which may remind us of a Roman Camp (*castra*) at that place. But the great majority of the place-names are either Celtic or Anglo-Saxon in origin, the older set, the Celtic, being attached to natural features or to places named from these, while the later English names are chiefly those of towns and villages. Thus the rivers Almond, Leith, and Esk have Celtic names; so have the hills, such as the Braids (G. *Braad*, a breast), Calton (G. *Calltuim*, hazels), Torphin (G. *Torr*, a round steep hill), and Drumsheugh, formerly Drumselch (G. *Drum sealg*, the ridge of hunting), which was the name of the forest that stretched south of Edinburgh. Celtic also are such old village names as Cramond (G. *Caer Amuin*, the stronghold on the river), Roslin (the headland of the pool), Currie (*curragh*, a marsh), and Ratho (the place of the forts).

The numerous village and farm names ending in *ton* are obviously of English origin. Granton is the *grene tun*: Liberton is the town of the lepers, so named as far back as Malcolm Canmore's time and long resorted to on account of the healing properties of the "oylie-well" or "balm-well" of Saint Catherine: Newbattle is *niwe botl*, the new house: Stow is the place; it was once called the Stow of Wedale and is still spoken of locally as "the Stow."

Interesting are Burdiehouse (Bordeaux house), and Little France, as reminding us of the "auld alliance," the *entente cordiale* of an earlier day, between France and Scotland. They were named from being inhabited by Frenchmen in the days of Queen Mary. Picardy Place was originally Little Picardy, a bleachfield on the outskirts of Edinburgh, where, in the first quarter of the eighteenth century, French weavers, brought over by the Board of Manufactures from St Quentin, taught cambric weaving to their less skilled Scots brethren of the craft. Portobello is reminiscent of its prototype on the Spanish Main, at the taking of which in 1739 the founder of the later town was present.

At the census of 1901 Midlothian stood second in order of the counties of Scotland in respect of population. The numbers for the whole country were 4,472,043 and for the county 488,796, roughly one-ninth of the total. Midlothian had 1335 persons to the square mile as compared with 1523 for Lanarkshire, the most densely populated county, and 11 for Sutherland, the most sparsely peopled of all.

The curves on p. 55 show the relative rate of increase in the two former counties during last century. Between 1861 and 1881 there was for Midlothian an increase of

115,167: while as between 1881 and 1901 the increase was only 101,632: that is 13,535 fewer were added to the population in the latter period than in the former.

We may note that while the proportion of females to males over the whole country was 105·7 to 100, in

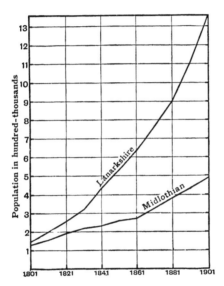

Curves showing the comparative growth of the populations
of Lanarkshire and Midlothian

Midlothian there were 112 females to every 100 males, the figures being 258,341 and 230,455.

The census tables permit of further interesting comparisons. Thus we are told that for all Scotland 3·8 per cent. of the population lived in one-roomed houses, in

Midlothian 3·54; for all Scotland there were 4·82 persons to each house, for Midlothian 4·85; for all Scotland there were on the average 3·26 rooms to each house, for Midlothian 3·71; for all Scotland there were 1·48 persons to each room, for Midlothian 1·31. These figures indicate that in Midlothian the number of persons in a household and the style of living are above the average for the country.

In the matter of resident foreigners Midlothian comes second to Lanarkshire but at a long interval. In all Scotland there were in 1901, 22,627 foreigners: 13,438 resided in Lanarkshire and 3456 in Midlothian. Of these the majority were Russians, Italians, and Germans. They have considerably increased in recent years.

A long and varied list of occupations is given for the inhabitants of the county. The professions were followed by over 8000 men and 5000 women; and 5000 men were engaged in national and local service. Agriculture claimed 5000 men and 1500 women. Engineering, working in metals, building, paper-making, and printing were each very important as trades for men. As in all urban districts, commerce, transport, provision and distribution of food, drink, tobacco, etc. afford work for large numbers of the population; and many men and women find employment in the making and selling of dress fabrics. Domestic service was the chief occupation for women, of whom 30,000 were so engaged, as compared with 3000 men. Doubtless also many of the 134,640 women returned as without specified occupation might be considered as busied with household duties. But the fact that 31,176 men were also returned as without specified occupation points to the popularity of Edinburgh with those who have retired from active work.

11. Agriculture.

The Lothian farmer has gained wide reputation as a skilled and enterprising agriculturist; and the plains of the Lothians have been famed for at least a century as one of the most highly cultivated districts in the world. But previous to the Union of the Parliaments of England and Scotland in 1707, matters were very different: the country was poor, the farms were mostly mere crofts, and little progress in method had been made for hundreds of years before. The old "runrig" or open-field system of culture was largely in vogue, while draining and manuring were little practised. The land was rested and fertility so far maintained by one-third or one-fourth of the arable land being left out of cultivation annually. The tenure was mainly that known as "steelbow," in which the farmer rented from the landlord not only land but also stock and implements and paid his rent largely if not altogether in kind. The implements were of the rudest sort: corn was thrashed with the flail and winnowed in the open air or in a draught between two doors. Farm animals were poor in quality; and, on account of the scarcity of fodder, as few as possible were kept over winter.

The impetus to improvement was due largely to the action of public-spirited landlords in the eighteenth century.

Sir John Dick of Prestonfield, when Lord Provost of Edinburgh, contracted to remove at his own expense the street refuse from the city, and by applying it to his land at the southern base of Arthur's Seat he converted an unprofitable marsh into highly productive fields. John, Earl of Stair, and Charles, Earl of Hopetoun, were great experimenters, and under their management a regular system of rotation in cropping was introduced; new crops were brought into the range of field culture, such as turnips, carrots, and cabbages; and the sowing of grass and clover became general. Later pioneers carried out draining on an extensive scale, practised deep culture with the new swing plough, applied lime and artificial manures to the land, adopted the thrashing mill, and otherwise improved their machinery and implements. Farm animals were more carefully bred, while the abundant supply of fodder made the wintering of stock and the production of beef and mutton profitable ventures.

Agriculturally, Midlothian may be divided into the coastal plains towards the north and the higher district further inland, rising into the hill country of the south. The coastal plains contain the land most valuable for arable purposes, the soil being in many places a deep rich loam, and the climate comparatively mild and dry. Except in the river valleys the southern hill region has poorer land and a moister, colder climate. The hill-farms are accordingly mainly pastoral, some of the hills being moorish and mossy, others covered with a thin clay.

According to the latest Ordnance Survey revision the

total area of the county is 236,595 acres. Land alone measures 234,325 acres and of this 75,044 acres are mountain and heath land used for grazing; 45,681 are under permanent grass; 81,082 are arable; and 12,336 are woods and plantations. Thus 85·3 per cent. of the total area is used for agricultural purposes, as compared with 73·2 for the whole of Scotland : 5·2 per cent. is utilised as woodland, compared with 4·6 for the whole country.

Inasmuch as different crops require different conditions for their successful growth, variations of altitude, slope and exposure, temperature, rainfall, and the geological nature of the rocks underlying the soil, all have a very direct bearing on the kind and quality of the crops grown in different districts. Thus, for example, in Midlothian the cultivation of wheat is usually confined to places under 500 feet above sea-level if the slope is to the north, and under 700 feet if on a southern slope, while the average summer temperature must be about 56° F. It is noticeable too that less wheat is grown in the west, where the rainfall is greater.

Fertility of soil requires to be maintained by varying the crop from year to year. The same crop naturally extracts always the same soil constituents for food, and to prevent exhaustion of these, different crops are grown in succession. The rotation is varied to suit conditions of soil and climate. In the most highly farmed parts of Midlothian, what is known as the "six-course shift" is common—the order of rotation being—Oats, Potatoes [part Beans], Wheat, Turnips, Barley, Hay or Pasture.

Suppose the cropping in any year is as follows in six fields—

1 Oats	2 Potatoes Beans	3 Wheat	4 Turnips	5 Barley	6 Hay or Pasture

In the second year oats will be grown on No. 6 and the other crops will be each shifted one place up the scale. Thus it is the seventh year before the same crop is grown on the same field. Even so, this is an exhausting use of the ground and the land is consequently heavily manured, especially with the potato turn.

Another rotation used where cattle are fed is the "five-course shift"—(1) Oats, (2) Potatoes and Turnips, (3) Barley or Wheat, (4) Hay, (5) Pasture; and where the land is a strong clay the rotation is frequently (1) Oats, (2) Beans, (3) Wheat, (4) Turnips, (5) Barley, (6) Grass.

In 1909 the acreage under the chief crops was as follows:

Wheat	...	...	...	5832 acres
Barley	...	...	...	4882 „
Oats ...	...	...	...	20,518 „
Potatoes	...	...	...	7120 „
Turnips	...	...	...	10,368 „
Clover, etc. (in rotation)		...		29,961 „

Wheat is usually sown after roots or beans, the seed being put in either in the late autumn or in the early spring. The crop averages 46 bushels per acre, as compared with 39 for all Scotland and 32 for England.

Barley occupies on the average about one-eleventh of the area under crop and gives a return of from 42 up to as much as 60 bushels per acre; the average being about 43½ as compared with 35 for Scotland and 33 for England.

Oats usually come after lea and are sown about the beginning of March on the low ground and some three weeks later on the higher levels, where it is the chief or only grain crop. The yield is about 42½ bushels; the figures for Scotland and England are 36½ and 41½, oats being the one cereal in which the English average beats the Scottish.

Beans need deep tilth and plenty of lime in the soil. They give a return of 32 to 36 bushels: but the area under this crop is very small, reaching in 1909 only 16 acres.

For hay and grass the seeds are usually sown with the barley. Much of the grass near the city is cut green for feeding the cows of city dairymen, who buy the growing crop and usually take two or three cuts in a season.

Turnips occupy almost one-tenth of the cultivated area, being now very important as winter food for cattle and sheep. Sixteen to 18 tons per acre for swedes and 22 to 23 tons for common turnips are considered good crops.

The soil and climate of Midlothian are favourable to the cultivation of potatoes, of which over 7000 acres are grown. This crop is an expensive one as the land requires to be well wrought and heavily manured. Seven to 8 tons per acre is a fair return. Potato culture is peculiarly interesting in the county, for it was near Edinburgh that Henry Prentice, a city pedlar, about the middle of the eighteenth century first grew potatoes as a field crop.

Many growers now adopt the plan of "boxing" the tubers selected for sets, in order that growth may be commenced before planting, as by this means an earlier crop and better prices may be obtained.

Other crops grown are cabbage, vetches, mangolds, and carrots.

In view of what has been said as to the skill of the Lothian farmer it may be interesting to note, that Midlothian in respect of yield per acre takes first place among Scottish counties for wheat, barley, and hay, third for potatoes, fourth for turnips, and fifth for oats.

Cattle are bred on the higher farms in the county and a good deal of winter feeding for beef is carried on in all parts, the animals used being mostly strong shorthorn crosses from the south of Scotland, the north of England, and Ireland. Dairying is important, especially on the outskirts of the city. The county has about 20,000 cattle of all kinds.

Midlothian is the fourteenth county of Scotland in order of number of sheep kept. On the hills the small and hardy Blackfaced are practically the only breed, with some Cheviots on the lower farms; while in the plains half and three-part bred sheep, chiefly crosses of Border-Leicester and Cheviot, are the favourite stock. Shropshires, Oxfords, and Dorsets are kept in small numbers for crossing. The total number of sheep is over 190,000.

Little horse-breeding is carried on, but the stock of farm-horses—mostly of the Clydesdale type—is exceptionally good. Of these there are about 4500 in the county.

Pigs to the number of nearly 10,000 are bred and fed

chiefly in the neighbourhood of Edinburgh. Poultry are kept at most of the farms.

The average size of holding in the county is about 200 acres and most farms are held on a lease of nineteen years with breaks at shorter intervals. There are in all 1024 holdings for agricultural purposes and of these 155 are owned by the occupiers. The rental varies from a few shillings per acre on hill farms to £5 or even more on more fertile lands and nearer the town. The buildings are generally substantial, modern, and well designed, and fences and drains are maintained in good order.

Much of the best land is occupied by nurseries, market gardens, and orchards, which have a ready sale for their produce in the city. The districts immediately to the east near Musselburgh and to the west about Corstorphine and Ratho are specially noted in this respect. In 1909, 463 acres were returned as under carrots, onions, and other crops of similar character; 244½ acres as under small fruit; and 65 acres as orchard.

The parks and pleasure grounds surrounding the numerous mansions are in many cases finely wooded, as are also some of the hill slopes, 12,336 acres being the extent of woodland according to the latest returns; that is about one-nineteenth of the whole area.

12. Industries and Manufactures.

While it cannot be said of Midlothian that it is a great industrial district like the West Riding of Yorkshire or the lower ward of Lanarkshire, it can boast, nevertheless, of a remarkable variety of trades and occupations carried on within its bounds. Some of these are noticed separately under Agriculture, Mines and Minerals, Shipping, and Fisheries.

The rise and the growth of several of the industries are directly traceable to certain physical conditions. For instance, the geological structure of the district has resulted in the presence of abundance of good building stone, of a useful clay, and of a very fine limestone. Edinburgh itself is largely built of grey freestone quarried in the immediate neighbourhood. Craigleith Quarry on the western outskirts supplied most of the material used in building the New Town, and the durable character of the stone is evidenced by the state of the houses after standing for nearly a century. The Old Red Sandstone of Craigmillar and the white and blue stones from Hailes have been extensively used in the southern and western suburbs. These quarries are now less resorted to, but others have been opened near Bonnyrigg, East Houses, Gilmerton, Penicuik, and elsewhere. There are also several quarries for the supply of paving material and road metal at Blackford Hill, Esperston, Catcune, Middleton, Fountainside, Silverburn, and Ratho.

An extensive clay field at Portobello is probably to be regarded as part of the 100-feet raised beach deposits: and it has long been drawn upon for the manufacture of bricks, tiles, and pottery. Bricks and fire-clay goods are made at several of the collieries, including Whitehill, Arniston, Newbattle, and Niddrie, where the material used is either a soft shale known as "blaes" or the clay underlying the coal.

Burdiehouse Limestone has been famous for more than a hundred years, and now the quarries, as such, have had to be closed on account of the depth to which they had reached and the great amount of the overburden. The stone is still obtained by means of a mine at Burdiehouse and a shaft, 50 fathoms deep, at Straiton. The same limestone is also mined at Newpark, East Calder, and Harburn. At Cousland, limestone is quarried and finely ground along with shale in order to produce Portland cement.

At the present time 36 quarries are at work in the county and these employ altogether about 600 men. They produced in 1909, 50,692 tons of clay, brick-earth, marl, and shale; 30,204 of gravel and sand; 31,244 of limestone; 26,361 of sandstone; and 133,986 of other rocks and minerals.

The breweries of Edinburgh are likewise connected with geological conditions. They are dependent upon a constant supply of permanently hard water of a peculiar quality, and this water is obtained from the strata of the Upper Old Red Sandstone as developed in the southern district of the city and lapping round Blackford Hill towards Craigmillar. These strata dip to the north, north-west, and north-east, and conduct the water in a northerly direction underneath the overlying cementstone group until it reaches the Colinton and Calton faults, which act as dams and impound it. At Duddingston the Craigmillar sandstone at the top of the Old Red also gives a plentiful supply, and in recent years several breweries have sprung up there. Water obtained from sources beyond the limits so defined appears to be lacking in some of the necessary ingredients or to possess others less desirable, and so is less suitable for brewing purposes.

There are breweries also at Musselburgh and Dalkeith. Distilling and the rectification of spirits are extensively carried on in the city.

Brewing and distilling are dependent on supplies of barley, and doubtless to begin with, these would be obtained mainly from the surrounding district.

The rich agricultural neighbourhood was responsible also for the rise of the great milling and baking industries of Edinburgh and Leith. The Water of Leith has long been noted for the mills which borrowed their motive power from the stream, and the great flour-mills at Haymarket, Stockbridge, and Leith, though no longer depending on the water wheel, are the lineal descendants of older-fashioned establishments on the same sites. In recent years biscuit-making has become a very important branch of the Edinburgh baking trade.

As the metropolis of the kingdom of Scotland, the chief place of residence of the sovereign, and the seat of the supreme courts of law, Edinburgh was naturally the first town to possess a printing press. Only thirty years after Caxton introduced the art into England, Walter Chepman and Andro Myllar in 1507 set up a press in the Cowgate at the foot of Blackfriar's Wynd. In a patent granted to these burgesses by King James IV it is set forth that "they at his Majesty's request, for his pleasure, and the honour and profit of his realm and lieges, had taken upon them to bring hame ane print, with all stuff belonging thereto and expert men to use the same for imprinting within the realm of the books of the laws, Acts of Parliament, chronicles, massbooks and portuns after the use of the realm with additions and legends of Scottish saints now gathered to be eked thereto and all other books that shall be necessary: and to sell the same for competent prices by his Majesty's advice and discretion, their labour and expenses being considered." The first book known to have been printed by Chepman and Myllar is a volume of metrical tales and ballads. Printing, so introduced, remained a struggling industry for nearly two hundred years and even when Bibles, "weel and sufficiently bund in paste or timmer" were issued at the price of £4. 13s. 4d. each, the Council had to enforce their purchase by enacting that all persons worth £500 should possess a copy or incur a penalty of £10. But about the middle of the eighteenth century the industry began to develop rapidly: Edinburgh became a noted literary centre, and soon gained for her printing a world-wide reputation which she has ever since retained. The names of Creech, Constable, Ballantyne, Blackwood, Black, Chambers, and Nelson are known wherever English books are read; and the *Waverley Novels*, the *Encyclopaedia Britannica*, the *Edinburgh Review*, *Blackwood's Magazine*, and *Chambers's Journal* have spread the fame of Edinburgh for enterprise in printing and publishing. Stereotyping, a process which has revolutionised the production of printed matter, was the invention of an Edinburgh silversmith, named John Ged. Other branches of the industry, such as engraving, lithographing, and mapmaking are all carried on with much success. Subsidiary industries have also arisen, and large numbers of work-people find employment in the making of printing machinery, typefounding, the manufacture of ink, and bookbinding.

The most important of these dependent trades is the manufacture of paper—one of the leading industries of the county. The valleys of the Esk and the Water of Leith contain numerous paper mills, the oldest being that at Valleyfield, Penicuik, which was established in 1709 by Anderson, the official printer to Queen Anne. In the Esk Valley there are mills at Penicuik, Glencorse, Springfield, Polton, Lasswade, and Musselburgh; in the Leith Valley, at Balerno, Currie, Juniper Green, Colinton, and Slateford. Much of the raw material used in the process, such as linen rags from Central Europe, esparto grass from North Africa, and wood pulp from Scandinavia, is landed at Leith.

Apart from the industries above referred to or noted separately, Midlothian has no great staple of production; but, especially in the city, there is a very considerable diversity of arts and crafts, of which a few of the more interesting may be mentioned.

The making of rubber goods was introduced in 1855, and in spite of keen foreign competition a large output is maintained of water-proofed garments, overshoes, and motor and bicycle tyres. The special skill of the Edinburgh artist in gold and silver work, electro-plating, and jewellery-making has long been widely known; and, though little is now done in the actual making of watches and clocks, about 2000 men are employed in various ways in this particular branch of industry. The making and the repairing of scientific apparatus of all kinds, specialised medical appliances, and optical instruments, are necessarily important in a town where education and scientific research are highly organised and which boasts of one of the most famous medical schools in the world. Similarly, several firms are engaged in the manufacture and supply of chemicals.

Textile manufactures in the county are not specially noteworthy; but there are carpet factories at Bonnyrigg, Dalkeith, and Roslin; while hosiery and homespuns are produced at Restalrig. Stow is in the Tweed basin, and there we find a factory of the characteristic woollen cloths —the *tweels*, misnamed tweeds.

There are numerous engineering works in Edinburgh and Leith, small iron-foundries at Slateford, Millerhill, Musselburgh, Loanhead, and Penicuik, and brass-foundries in Edinburgh. Gunpowder is made at Roslin, Camilty (Midcalder), and Kirkettle. Rope-works are carried on at Leith, Dalkeith, and Musselburgh. At Musselburgh, also, nets are woven, and twine and wire manufactured. Within the last few years the wire works have been exceptionally busy in supplying the demand for wires used in the construction of aeroplanes.

Glass bottles are made at Portobello and fine-cut glass or crystal is produced in considerable quantity and in the best style by the well-known works near Holyrood.

Gelatine and glue are manufactured at Gorgie and Cramond. Several tanneries carry on operations in the city and in Leith, where also there are large factories of chemical manures and cattle-feeding stuffs. Leaf tobacco is imported at Leith and Granton, where there are large bonded stores, and is manufactured in several city factories. Two small mills at Juniper Green are still engaged in the making of snuff; but the mill in which James Gillespie, the founder of James Gillespie's Hospital, carried on his business in the second half of the eighteenth century, no longer exists.

Salt works on a small scale continue in operation at Joppa Pans and Pinkie Pans between Portobello and Musselburgh, cargoes of salt rock being landed at Fisherrow Harbour.

The numerous golf courses in the vicinity of the capital give employment to a goodly number of makers of golf clubs and rubber-cored balls as well as to the green-keepers and "caddies."

It has been often remarked that one of the chief industries of Edinburgh is education. Few cities are so well supplied with the means of instruction—admirably equipped and staffed Elementary, Higher Grade, and Secondary Schools, various highly efficient Technical Institutions, and an ancient University. The fame of these educational establishments attracts great numbers of students from all parts of the world; and the University of Edinburgh is possibly more cosmopolitan than any other in the varied nationality of its undergraduates. The reputation of the city as a place of learning and the existence of several great libraries, access to whose treasures is readily obtainable, together with its educational facilities, make it a favourite place of residence for wealthy and leisured people with intellectual tastes.   Professional men form a very considerable portion of the community.   A large part of the legal and factorial business of the country is transacted in the city; and the men of law—agents, solicitors, writers to the signet, advocates, and judges— are a numerous and influential body.   Edinburgh has long been recognised as a training centre for accountancy, in- surance, and banking; and many important banks and insurance companies have here their headquarters.

13. Mines and Minerals.

To the monks of Newbattle probably belongs the credit of being the first discoverers and the first workers of minerals in Scotland.   They are said to have worked coal beds near Prestonpans as early as the thirteenth century.   In a charter granted to them by Sieur de Quinci, Earl of Winton, between 1210 and 1219, mention is made of a coal heugh between Pinkie and Prestonpans.   This heugh was probably a day-level driven into the seam from the outcrop near the sea.   In 1526 the monks formed, near the boundary of the two counties, a harbour, now known as Morison's Haven, for the export of the coal obtained from their mines.

In the Lasswade district coal appears to have been worked at least as early as the beginning of the sixteenth century.   But the total output must have remained small until well on in the eighteenth century.   Demand was small.   In country districts, peat, and in towns, timber were the regular fuels; there were no great industries to call for large supplies; and the means of conveyance for coal were of the most elementary type—panniers on the backs of horses travelling along the wretched roads at snail's pace. Moreover, the methods of production were rude and clumsy, involving human labour under peculiarly trying conditions. So hard indeed was the life of the

Newbattle Abbey

collier that an Act of the Scots Parliament had to be passed in 1606, which practically enslaved the workers in mines as "necessary servants," who, if they left their masters, might be proceeded against as thieves for stealing themselves and their services. They became a hereditary caste or class apart; for if the sons or daughters of colliers once engaged in the work they thereupon became serfs themselves for life. Children also were often sold by their parents into this condition. Little wonder that in course of generations was developed a characteristic type of degraded humanity, marked, as Hugh Miller describes the collier women of Niddrie, "by a peculiar type of mouth, wide-open, thick-lipped, projecting equal above and below, resembling that which we find in prints of savages in their lowest and most depraved state." It was not until the close of the eighteenth century that complete emancipation was given by legislation to the colliers of Scotland.

The only minerals of any importance obtained in Midlothian are coal, ironstone, and oil-shale. In 1909 there were 30 mines at work and they gave employment to 7892 persons below ground and 2040 above, a total of close upon 10,000 employees.

Geologically, these minerals are obtained from the carboniferous strata, which in this district occur immediately under the boulder clay. They are arranged as under :

Carboniferous Strata			
	Middle Coal Measures		
	Lower Coal Measures		
	Upper Roslin Sandstone or Millstone Grit		
	Lower Roslin Sandstone or Millstone Grit		
	Carboniferous Limestone	Upper Limestone group	
		Edge Coal group with Ironstone	
		Lower Limestone group	
	Calciferous Sandstone	Oil-shale group	
		Cementstone group.	

It will be observed that the oil-shales are got from a geologically older formation than the coal, and that the ironstone occurs along with the coal of the Edge Coal group. In Midlothian this group is much more important than the Coal Measures of later formation, which like corresponding strata elsewhere are named the Lower and Middle Coal Measures.

The Midlothian Coalfield lies in a basin bounded on the west by a great fault which runs in a N.N.E. direction parallel with and on the eastern side of the Pentland Hills. On the east the coal-bearing strata rise in an anticlinal fold about Roman Camp Hill and dip again into the basin which forms the East Lothian Coalfield. The Midlothian basin or trough appears to have been shaped by pressure acting from the west, for the strata near the Pentland fault are inclined at a very steep angle, which in some places even goes beyond the perpendicular so that the rock positions are reversed: whereas in the east of the basin the slope is much more gentle. In the middle part of the basin the seams are flat.

On the north, the coal-bearing strata pass under the Forth to re-appear in the Fife coalfield; and they extend from the coast southwards into the heart of the county beyond Rosewell, a distance of about nine miles. The strata are thickest near Niddrie in the north-west part of the basin, and thin out towards the south and east. At the most fully developed position, the Edge Coal group is fully 1000 feet thick and contains 26 seams of coal of more than one foot thickness. On the western margin of the basin they may be followed from the sea at Joppa by Wester Duddingston, Niddrie, Drum, and Gilmerton to Loanhead and Dryden. They are called " edge " coals locally because the seams present their edges upwards instead of lying in their original "flat" or horizontal position. The eastern margin is to be found at Wallyford, Carberry, Cowden, Newtongrange, and Gorebridge ; and there, as has been noted, the dip is at a much smaller angle, while in the intervening region the seams gradually reach the horizontal at a great depth, from 2756 to 3500 feet. At Niddrie, coal is raised from a greater depth—2623 feet vertical—than anywhere else in Scotland.

The lie of the strata and the depth to which the seams have to be followed make the winning of the Edge coal a difficult and expensive process. As compared with a pit sunk to " flat " seams, one sunk to " edge " coals opens up only a small area of workable coal ; and it is necessary therefore to drive at great expense cross-cut mines through the intervening strata to connect the various seams and permit of the coal being drawn up the same shaft to the surface. Some of the shafts are sunk in the direction of the dip and the coal is hauled to the surface on rails laid on the "pavement" or rock underlying the coal. On account of the cost, only the best and thickest seams are worked at present.

What have been called the Lower and Middle Coal Measures lie above the Edge Coal group in stratigraphical order. Of these Measures, the Middle—about 500 feet thick—occupies an area of less than a square mile near the middle of the district and contains no workable coal seams. The Lower also lies in the middle of the basin and contains several useful seams grouped in two divisions with 200 feet of barren strata separating them.

In recent years the Midlothian Coalfield has been considerably developed. In 1903 the output of coal in the county amounted to 1,623,390 tons: in 1909 2,634,355 tons were produced, an increase of close upon a million tons. This increase has taken place in spite of the fact that the value of some of the coals has fallen very much on account chiefly of newer methods of gas production. Gas coal was at one time chiefly used for that purpose and brought a high price; but now owing to the introduction of incandescent mantles, gas of a lower illuminative power satisfies the demand and ordinary coal is used. Gas coal was nicknamed " Cannel " or Candle coal from the high illuminative power of the gas it produced in burning, and " Parrot " coal because of the chattering noise it makes in the fire.

In the Report of the Royal Commission on Coal Supplies (1905) it is estimated that the quantity of available coal in the Lothian Coalfield—including Haddingtonshire and Peeblesshire—may be put at 2,520,311,573 tons, of which 2,365,000,000 tons may be considered as belonging to the Midlothian field. If to this be added one-third of the supply lying beneath the Forth, which might be reached from the southern shore, it is calculated that the present output could be maintained for nearly two thousand years.

The principal collieries are at Olive Bank, New Craighall, Niddrie, and Woolmet; Whitehill and Polton; Loanhead and Roslin; Penicuik; Arniston, Vogrie, and Newbattle; Carberry and Wallyford.

At Niddrie, Loanhead, and Penicuik, blackband and

clayband ironstone are worked along with the coal, and the mineral is sent to Shotts to be smelted. The total output of ironstone in 1903 was 21,775 tons; in 1909 37,772 tons.

The neighbouring county of West Lothian was the scene in 1851 of the first development of the paraffin oil industry, when James Young erected retorts and refineries at Bathgate to obtain oil from the famous Torbanehill Mineral or Boghead coal, which was so rich in oil that its price rose to £4. 10s. a ton before it was exhausted.

Shale was first used for the distillation of oil at Broxburn in 1859 and now it is the only mineral employed for that purpose. The shale is produced in Midlothian at Cobbinshaw, round about West Calder, and at Pumpherston; formerly also at Straiton. In 1909, 638,915 tons of oil-shale were raised.

Retorting works and refineries are in operation at Pumpherston, Oakbank, and Addiewell.

In the process of extracting the oil the shale is first broken and distilled in retorts at a temperature of from 600° to 800° F. This distillation yields crude oil and ammonia water. The crude oil is refined by further distillations, by chemical treatment, and by cooling and pressing. From it are obtained naphtha, various grades of burning and lubricating oils, and solid paraffin or wax. The wax is made into candles. From the ammonia water, which at first was merely allowed to run off as waste, there is now got a substance of considerable commercial value—sulphate of ammonia—which is largely used as an artificial manure.

14. Fisheries.

Not the least of the advantages which we owe to our insular position off the coast of Europe is our proximity to the feeding grounds of enormous numbers of edible fish which live in the shallow waters around these islands. Fishing has always been a leading industry of Scotland, and fish has been a staple of export for centuries.　An

Newhaven

indication of the importance of the industry is given in the statement that the average annual value for the 21 years ending in 1908 of all the fish landed at Scottish ports was £2,047,316, while the number of persons employed in connection with the various branches of the industry was 92,837.

The adaptation of the steam-engine to sea-going vessels added greatly to the facilities for capturing fish and

to the ease and rapidity with which they can be brought
to market. While therefore the picturesque two-masted
yawl with its brown lug-sails and its two-mile-long lines
of hooks baited with mussels still goes out from Newhaven
or Fisherrow to fish in the waters of the Firth—preserved
to the line fishermen since 1887—the bulk of the fish is
now landed at Granton or Leith from trawlers, which
either work on the grounds extending from 20 to 60
miles E.S.E. and E.N.E. of the May Island or bring
their week's catch from the more distant grounds of the
Buchan Deeps, or off the Fair Isle, or around the Orkney
and Shetland Islands.

Line-fishing, however, has the advantage over trawling
that it can be used where the sea-bottom is rough and
rocky. An interesting development of line-fishing has
taken place in recent years, notably on the East Coast, in
that steam-vessels have been introduced; and the number
of these has increased from 38 in 1899, to 337 in 1908.
Steam-drifters, too, are being more and more employed in
net-fishing; and the number of sail-boats is steadily declining.

In the Leith district, which includes all creeks from
Cove in Berwickshire to Dysart and Wemyss in Fife, the
number and values of the vessels employed in 1908 together with the number of men engaged were as follows:

| Vessels | | Value | | No. of |
		Boats	Gear	Fishermen
Sail-boats	382	£38,004	£38,016	1461
Steam-drifters	7⎫	14,700	3,695	90
Steam-line boats	3⎭			
Steam-trawlers	61	232,000	8,540	549
Totals	453	284,704	50,251	2100

The principal fish caught by line and trawl are such as feed on or near the sea-bottom in comparatively shallow waters. They are classed as *demersal* to distinguish them from fish living near the surface and known as *pelagic*. The former class is subdivided into flat-fish such as halibut, turbot, brill, soles, plaice, flounders, dabs, lemon soles, skates, and rays: and round-fish such as cod, haddock, whiting, hake, ling, gurnard, red mullet, bass, and dory. Of pelagic fish the chief are herring, mackerel, pilchards, sprats, and sparlings; and of these by far the most important in Scotland is the herring, which is regarded as the stand-by of the Scots fisherman. Herrings are caught by drift-nets mainly: but the Newhaven fishermen in the upper reaches of the Firth use the seine net, which is shot in a semi-circle, then has its ends drawn together, and thus encloses its prey; whereas in the drift-net the herrings are held in the meshes by their gills.

In 1908 the quantities and values of fish landed at Leith and adjacent ports were:

Herrings	27,335 cwts.	value £5,707
Sprats, Sparlings and Mackerel	5,784 ,,	,, 1,088
Round Fish	268,240 ,,	,, 121,816
Flat Fish	35,882 ,,	,, 45,442
Other Fish	9,213 ,,	,, 2,669
Total	346,454 ,,	,, 176,722

In addition shell-fish—oysters, mussels, clams, lobsters, crabs, etc.—were landed to the value of £5625, making the aggregate value of the fish of all kinds caught during the year £182,347.

The local herring fishing is carried on in the winter months, and in recent years the catch has been disappointing. Many of the larger boats take part in the summer fishing chiefly at other east coast centres from Fraserburgh to Yarmouth; and in this way, in 1908, 34 boats supplemented by as much as £11,890 the income obtained in home waters.

On account of its position Leith is a convenient centre for the dispatch of cured fish to continental ports; and in 1908 the number of barrels of herrings exported was 217,504 and of other fish 17,266. So far as herrings are concerned, however, there is a growing tendency to ship them direct from the curing centres. The chief foreign ports to which the fish from Leith were consigned were St Petersburg, Riga, Königsberg, Danzig, Stettin, Hamburg, and Rotterdam.

Considerable quantities of shell-fish chiefly crabs and clams with some lobsters and oysters are landed at Leith, only the Orkney, Skye, and Ballantrae districts having a greater total. Oysters are part of the stock-in-trade of the Newhaven or Musselburgh fish-wives, and their musical cry "Caller Ou" was at one time as familiar in the streets of Edinburgh as are the sturdy figures of the women themselves, who in their quaint and picturesque costumes and with their well-laden creels on their backs bring to the doors of the citizens "the bonnie fish— the halesome farin, new-drawn frae the Forth."

Dependent on the main industry are numerous subsidiary employments, such as bait-gathering, net-making and mending, boat-building, gutting and packing fish,

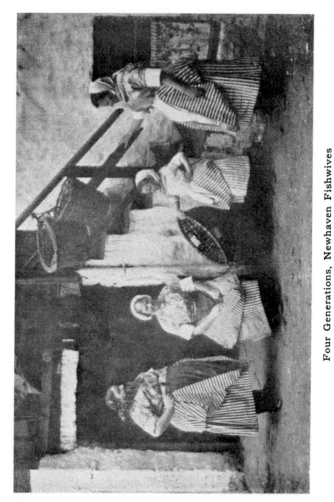

Four Generations, Newhaven Fishwives

curing, making barrels, baskets, and boxes, hawking and selling fish and so forth. It is calculated that in the Leith district these subsidiary industries give employment to over 3000 people in addition to the 2100 men and boys actually engaged in fishing.

All matters relating to the fishing industry in Scotland are under the care of the Fishery Board, whose duties include the policing of the home waters to prevent illegal trawling and other abuses, the administration of the laws made from time to time to regulate the industry, the registration of the boats engaged, the branding of cured herring, the granting of loans to fishermen, the carrying out of scientific investigation of fishery problems, and the collection and collation of statistics.

15. Shipping and Trade.

The position of the Firth of Forth in relation to the Continent has made its shores the natural starting places for communication with the lands beyond the North Sea; and it is probable that harbours have existed at the mouths of Midlothian's three rivers from the earliest times. Cramond, at the mouth of the Almond, was important in the days of the Romans: the harbour of Leith, or Inverlet, as it was once called, was originally merely the channel of the Water of Leith worn out of the broad expanse of Leith Sands; and the oldest name for Musselburgh, Eskmuthe, recalls the fact that, though the modern harbour of Fisherrow is half a mile to the west, the port in earlier times was simply the mouth of the Esk.

The great development of trade in recent times necessitated a like development in the facilities for dealing with the more numerous and larger ships and the vastly increased traffic. The old creeks either decayed in importance or were obliged to extend their accommodation. Cramond has no longer any pretensions as a port, and the

Leith Piers

overseas trade of Musselburgh has wellnigh reached the vanishing point. Leith, on the other hand, as the gateway of the metropolis, has grown into a first rate seaport, of all Scottish ports second only to Glasgow; and two and a quarter miles to the west, Granton Harbour was constructed about the middle of last century with the special purpose of meeting modern requirements in loading and discharging cargo at all states of the tide.

The first attempt to improve the harbour of Leith was the construction of a pier in 1544 by the English general, the Earl of Hertford, who was then holding the port: but he destroyed the pier when he left for England. Various other quays and piers were constructed in the seventeenth and eighteenth centuries, but the bar at the river mouth was still frequently impassable. Between 1800 and 1817 wet and graving docks were made; while in 1824 the east pier was extended and the west pier and breakwater built. Since then the Victoria Dock, the Albert Dock, the Edinburgh Dock, and the Imperial Dock have been added. The total area of the harbour is about 100 acres. The entrance channel—3000 feet long and 231 feet wide between the piers—has a depth of 10 feet over the bar at low water of spring tides. The channel is not used for ships discharging cargo but is available for landing passengers. There are now eight dry docks for the overhauling of ships. The North British and Caledonian Railways have sidings on the quays; and railway trucks can be run alongside all the ship-berths. The docks are fully equipped with storage sheds, steam, electric, and hydraulic cranes, hydraulic coal-hoists, capstans, grain elevators, and other appliances, and there are large yards for the stacking of timber.

A considerable coasting trade is carried on with other British ports; and foreign trade is chiefly with Russia, Sweden, Norway, Denmark, Germany, the Netherlands, Belgium, France, Spain, and the United States of America. The following British Possessions are also represented: Cyprus, India, Australia, and Canada. Lines of steamers make regular passages to Newcastle, Hull, and London; to Dundee, Aberdeen, Orkney, and Shetland; and to continental ports on the North Sea and the Baltic. In summer, excursion steamers carry numerous passengers on pleasure trips to Aberdour, Stirling, Elie, and other resorts.

In 1909, 218 vessels of 153,021 aggregate tonnage were registered as belonging to the port: while during the same year 220 sailing vessels and 2469 steamers entered and cleared in coastwise trade: 104 sailing and 1824 steam-vessels entered and 117 sailing and 1768 steam cleared for foreign countries and British possessions. Altogether 6398 vessels with an aggregate of 3,504,717 tons entered and cleared in the year mentioned.

Imports were mainly of grain, hemp, provisions, esparto grass, fibre, ores, timber, artificial manures, wines, sugar, and fruit; and the value of these in 1909 amounted to £12,883,890. Exports were valued at £5,377,188, including foreign and colonial goods transhipped to the value of £111,772; and the chief articles of export were iron, hardware, coal, canvas, machinery, chemicals, firebricks, manures, herrings, spirits and ales.

Shipbuilding has long been carried on. The first line of battleship—the *Fury*—ever built in Scotland was launched here. In 1909, eight steel steam-vessels were launched of a gross tonnage of 4559 tons.

The stretch of water between the town and the island of Inchkeith is known as Leith Roads, and in easterly gales it is frequently crowded with craft, which find there good anchorage. Inchkeith is strongly fortified as a defence to the harbour, which previous to 1878 was guarded by a Martello tower built on the Black Rocks east of the entrance at the time of the Napoleonic wars, when it is said the French emperor had in contemplation an attack on Edinburgh by a fleet which was to sail up the Water of Leith !

Formerly the Corporation of Edinburgh claimed property in the harbour and exercised jurisdiction over it. This authority originated in a grant made by King Robert the Bruce in 1329 of "ane right of the harbour and mills of Leith, with their appurtenances to the city of Edinburgh to have and to hold in all time coming for the yearly payment of 52 merks" [about £280 in modern money value]. An Act passed in 1838 vested the management in the Dock Commissioners, who are elected by various public bodies in Edinburgh and Leith.

From early times the Association of Mariners and Shipmasters at the port was accustomed to receive from all vessels belonging to it and from Scottish vessels of other ports visiting it, a sum of money called "prime gilt" which was applied to aid "poor old infirm and weak mariners." The association is known as Trinity House; and, although "prime gilt" was abolished in 1862, the investments of the House enable the charity still to be carried on. Trinity House is also the authority for the licensing of Leith pilots.

The harbour has frequently been the landing place of royalty. James I, the poet king, brought to Leith his Queen, Jane Beaufort, whose beauty he sang in the *Kingis Quair*. His son James II, the boy king, sailed from Leith to Stirling after his abduction from Edinburgh

Castle. The consorts of James II, James III, James V, and James VI all disembarked here on their arrival in Scotland. After her brief, bright reign in France, it was at Leith that

> "The lovely Mary once again
> Set foot upon her native plain,
> Kneeled on the pier with modest grace
> And turned to Heaven her beauteous face."

The ill-fated Darien Expedition, with its five frigates containing 1200 men and 300 gentlemen, and bearing its cargo of Scots woollen manufactures—heavy plaidings and blue bonnets included—to trade to the natives on the Spanish Main, sailed from Leith in 1698. In 1705 Captain Green and two of the crew of the English ship *Worcester* were hanged on Leith Sands for alleged piracy committed on the high seas in 1703—evidence this of the bitter feeling then prevalent against Englishmen. Leith Fort was commenced in 1779 as a battery of nine guns, set up to defend the harbour against the threatened raid of the Scots-American privateer, Paul Jones, who in that year scared the inhabitants of the Forth towns by appearing in the Firth with two of his vessels. The guns were not needed, a westerly gale having opportunely arisen and driven the privateersmen out to sea. In later times the port was visited by George IV, Queen Victoria, and Queen Alexandra.

Up till nearly the middle of the nineteenth century there was no deep sea harbour in the Firth of Forth, and this want moved the Duke of Buccleuch, superior of the place, to develop the harbour of Granton, which was

opened on the Coronation Day of Queen Victoria, who
soon after visited it. The harbour is formed by two
breakwaters 3170 and 3100 feet long and has a total area
of 129 acres, with an entrance 340 feet in width and
a depth at low water, spring tides, of 13 feet. A pier
200 feet wide runs out between the breakwaters for 1700
feet and there are various jetties and slips for the landing
of goods at all states of the tide. The steam cranage
accommodation is excellent and there are coal cranes and
tips which load each about 110 tons an hour. The
Caledonian and North British Railways run alongside, and
the dispatch of cargo is easy and quick.

The port is the headquarters of several lines of
steamers trading to Aberdeen and other northern ports, to
London, to Christiania, to Gothenburg, etc.; also of the
preventive and the fishery protection vessels for the district.
Sweden, Norway, Denmark, Russia, the Netherlands,
Belgium, France, Spain, and Italy are all traded with;
chief imports—timber, iron, grain, esparto, cement, china
clay, rosin, turpentine, bottles, wood pulp, flour, paper,
motor spirit, and carbide of calcium: chief exports—coal,
coke, iron, oil, machinery, castings, whisky, and general
merchandise. In 1909 the numbers of vessels were :—

        Entering, Sail,   239 of 135,088 tons,
                  Steam, 315 of 238,132 tons,
        Clearing, Sail,   77 of 9324 tons,
                  Steam, 106 of 74,041 tons,

a total "movement" of 737 vessels of 456,585 tons.
Sixty-six vessels of 2635 tons were registered as belonging

to the port. The total value of the imports amounted to £748,707; of exports, being produce of the United Kingdom, £235,574, and exports, produce of foreign countries and British possessions re-exported, £3779: a total value of £239,353.

Before the building of the Forth Bridge the ferry from Granton to Burntisland was an important part of the North British Railway trunk line to the north, and an arrangement was in use by which loaded trucks were shipped on large steamers, conveyed across the Firth and run ashore on the other side, the loading, unloading, and reloading of goods being thus avoided.

16. History.

Properly to write the history of Midlothian it would be necessary to give in some detail the main events of Scottish history; for the county town has been long recognised as the metropolis of Scotland; and many of the incidents which have marked the stages of national progress have been enacted within the walls of the grey old city. Here we can deal only with events which may be regarded as connected rather with the county proper.

The region of the Lothians—not however under that or any other name—first emerges in history when Tacitus describes Agricola's campaigns (80 and 81 A.D.) in the lands to the south of the Bodotria or Forth. It was Agricola's idea that here should be the terminus of the

Roman Empire, and hence his line of forts across the narrow neck between the Forth and the Clyde to curb the wild Caledonians of the north. With his soldier's

Edinburgh Castle, as it was before the siege of 1573

eye he would perceive the strategical value of the Castle Rock, upon which the Ottadeni most likely had already built a fort. Its name in British—perhaps then, certainly later—was Mynedh Agnedh, "the hill of the plain."

Subsequent corruption of this to Maiden's Castle—in the Latin of the time *castrum puellarum*—originated the tradition that the Rock was the home of the royal princesses. After the English settlement, Lothian became in the sixth century merged in the Anglian kingdom of Bernicia and then in Northumbria. Edwin of Northumbria gave his name to the fortress on the Rock—Edwin's burgh. The Gaelic Dunedin may be a translation of the English designation or it may mean "the fort on the slope."

It would be impossible to give in brief compass a satisfactory account of the fortunes of the Lothians during the centuries when the four kingdoms—of the Scots, the Picts, Strathclyde, Northumbria—were struggling for supremacy. The end came in 1018, the year in which Malcolm II, king of Alban, defeated the Northumbrians at the Battle of Carham and so made good his title to the Lothians, since which time the district has remained a recognised part of the kingdom of Scotland.

In 1303, during the War of Independence, a small force of 8000 Scots, under Sir Simon Fraser and John Comyn, defeated near Roslin 30,000 English under Sir John de Segrave and Sir Ralph de Manton, who by order of Edward I were devastating the country. The English were in three divisions and the Scots were fortunate in being able to deal with each division separately. Fordun the chronicler tells quaintly how after the Scots had beaten the first division and were sharing the spoil a second column appeared in battle array; the Scots on seeing it, slaughtered their prisoners and armed their own vassals with the spoils of the slain ; then, putting away their jaded horses and taking stronger ones, they fearlessly hastened to the fray. This experience was repeated when, after a second victory a third English

Roslin Castle

force came upon the scene. Then the Scots "by the power not of man but of God subdued their foes and gained a happy and gladsome victory."

On the Boroughmuir to the south of Edinburgh, now included within the city bounds, Sir Alexander Ramsay in 1336 came upon a body of English troops under Guy, Count of Namur, on their way to join the army of Edward III, then encamped at Perth in support of Edward Balliol the usurper. Ramsay drove the English into the town and compelled them to surrender.

The alliance between France and Scotland culminated in 1385 in a combined raid into England commanded by Sir John de Vienne and the Earl of Douglas. In revenge King Richard II invaded Scotland and devastated the country as far north as Edinburgh. There he spent five days destroying the town with the church of St Giles' and the Abbey of Holyrood, but failed to capture the castle.

The Boroughmuir was also the scene of the mustering of the army which James IV led into England to the disastrous defeat at Flodden in 1513. The Bore Stone, on which his flag-staff was fixed, is now built into the wall in front of Morningside Parish Church, not far from its original site.

In pursuance of his policy of breaking up the Scoto-French alliance and uniting Scotland to England by the marriage of Prince Edward and Queen Mary, Henry VIII in 1544 sent an army into Scotland to force the Scots into compliance with his wishes—"a rough wooing." Under the Earl of Hertford this force landed at Leith, burned Edinburgh, wrecked Holyrood Abbey, dismantled the castles of Craigmillar and Roslin, and laid waste the surrounding district.

The Bore Stone

After Henry VIII's death in 1547, his policy was continued by Hertford, now Duke of Somerset, who again led an English army into Scotland. The Regent Arran hastily gathered the Scottish forces and took up his position on the ridge overlooking Musselburgh. Somerset was supported by an English fleet, which lay

Musselburgh Bridges

off the mouth of the Esk; but as the Scots were strongly posted he in turn stationed his army on the rising ground, Falside Brae, to the east of Musselburgh, and waited the attack of the Scots. The latter unwisely left their advantageous position and, filing over the Esk by the old bridge, climbed the ridge of Inveresk, suffering meanwhile from the galling fire of the English fleet. The main battle took place on the plain that slopes to the base of Carberry Hill, and was sternly contested for several hours. At last a body of Highlanders, mistaking for retreat a tactical movement by a party of the Scots, took to flight and infected with panic the main body, which was chased by the English with blood and slaughter a space of five miles westward and four miles in breadth; "in all which space the dead bodies lay as thick as a man may meete cattell grasing in a full plenisht pasture." The Scots loss in this Battle of Pinkie is said to have amounted to 10,000 slain and 1500 prisoners, and the day was long afterwards known as "Black Saturday."

The same neighbourhood was the scene twenty years afterwards of Queen Mary's dramatic parting from the Earl of Bothwell and of her surrender to the Confederate Lords. On the very ridge, since called Mary's Mount, where Somerset's soldiers had entrenched themselves before the Battle of Pinkie, the opposing forces met; but Bothwell's men deserted in such numbers that he was obliged to seek safety in flight and Mary had to put herself into the power of the nobles. Brought into Edinburgh through a jeering crowd, she was next day carried to prison in Lochleven Castle.

In the troublous times after the execution of Charles I, the immediate neighbourhood of Edinburgh witnessed the first passes in the duel between Cromwell and Leslie, which ended at Dunbar. After the proclamation of Charles II by the Scots, Cromwell was sent north with an army. Leslie's plan of action was the traditional Scottish one of laying waste the country before the invaders, cutting off

Mary Queen of Scots

supplies, and so starving them out. He posted his troops between Edinburgh and Leith both to prevent Cromwell's getting provisions by sea and to keep him from taking Edinburgh. Cromwell was compelled to encamp outside the city on Galachlaw, an off-shoot of the Braid Hills, where his men suffered much from sickness. When he attempted to march to Queensferry, Leslie intercepted him, and on the 27th of August 1650 an artillery duel took place near Gogar, in a field which came to be known afterwards as the Flashes. Unable to dislodge Leslie, Cromwell withdrew first to the Braids, then to Mussel-burgh, where his troops encamped on the links and used Inveresk church as a barracks, and finally to Dunbar. There Leslie's fatal blunder—a repetition of that which gave the English the victory at Pinkie—led to the igno-minious "Dunbar Drove."

In 1666 the county was once more involved in strife. In the attempt of Charles II to force episcopacy upon an unwilling people, the severity of legislation and its harsh enforcement by the military led to reprisals by Covenanters in the south-west of Scotland. Emboldened by some small successes, 2000 of them marched from Lanark towards Edinburgh: but they met with so little encouragement that at Colinton they turned for home. Striking along the foot of the Pentlands by Dreghorn Castle and Swan-ston to the Biggar Road, they marched to Rullion Green, where they halted. General Dalziel had been in close pursuit; and, hearing of the altered route, he led his troopers across Currie Brig and by the old drove road up the Kirk Loan and through the Maiden Cleuch, and

Swanston

came upon the Westland men on the southward slopes of Turnhouse Hill. After a brief resistance they were overcome; fifty of the insurgents having been killed and about a hundred taken prisoners. A stone marks the site of the battle, whose story is picturesquely related in R. L. Stevenson's *Pentland Rising*.

When Prince Charles Edward marched south in 1745, he approached Edinburgh from Falkirk by the old Glasgow road. His arrival at Corstorphine caused so great consternation that the mere appearance of a few Highland gentlemen was enough to raise a panic among the Royalist cavalry posted to the west of the city, and they fled back over the Water of Leith to the city; the affair being known as the "Canter of Coltbrig."

The Prince took up his quarters at Gray's Mill near Slateford and there received a deputation of citizens, of whom he demanded the immediate surrender of the town. While negotiations were proceeding, Lochiel introduced 900 clansmen into the town through the city gate, which had been opened to allow the passage of the coach that had carried the deputation to Slateford. The same day the Prince left his quarters and, making a detour to the south of the city in order to be out of range of the guns of the castle—still held for King George—came by the King's Park to Holyrood Palace, where for a time he held court.

After his victory over Sir John Cope at Prestonpans he returned to Edinburgh and blockaded the castle, but withdrew the blockade to save the town from damage by the castle guns. His Highlanders encamped on the eastern slope of Arthur's Seat, near Duddingston, and for their depredations gained an unenviable reputation among the peaceful inhabitants of the neighbourhood. After a month spent in Edinburgh, disappointed with the lack of support obtained from the Lowlands of Scotland, Charles set out on his futile march into England.

Edinburgh Castle

From the scene of his final defeat at Culloden, his conqueror, the Duke of Cumberland, brought back to Edinburgh a number of the captured standards and had them publicly burned at the Cross, the Prince's own by the common hangman and the thirteen others by as many sweeps.

17. Antiquities.

Archaeologists divide the prehistoric period of man's occupation of a region into the Ages of Stone, Bronze, and Iron according to the material of which his weapons of war and the chase or his implements of industry were manufactured. The Stone Age is subdivided into the Palaeolithic or Old Stone Age, and the Neolithic or New Stone Age. In the former the weapons were rudely chipped into shape; in the latter they were smooth, polished, and better adapted to the end in view. Of Palaeolithic times no remains seem to exist in Scotland; and not many even of the Neolithic Age have been found in Midlothian; but stone weapons and implements such as discs, flint flakes, arrow-heads, scrapers, hammers, and axes have been unearthed at various times in different parts of the county, notably Corstorphine, the slopes of Arthur's Seat, Duddingston Loch, Leith, Ratho, East Calder, Pentland, Roslin, Penicuik, Eskhill, Carlops, and Bavelaw Moor.

If we try to picture the face of the county in those far-off days, we must imagine it as largely covered with marshland and forest, out of which rose the slopes and summits of the hills. Of the ancient Caledonian Forest a relic of some hundred acres still remains in the park of Dalkeith Palace: at one time—as the Forest of Drum-selch—it reached to the outskirts of Edinburgh and was for many generations a favourite hunting ground of the Scottish kings. When the forest and marsh extended far

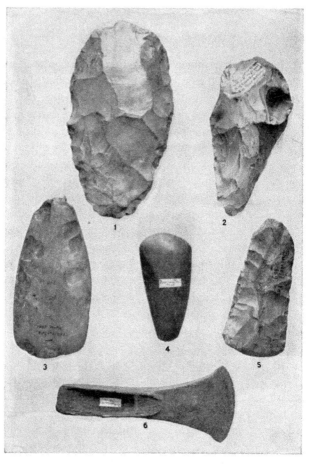

Weapons of the Stone and Bronze Ages

*1 and 2 are Palaeolithic; 3, 4 and 5 Neolithic;
6 is a Bronze Age palstave*

and wide the early inhabitants constructed their rude fortifications on hill tops or on artificial islands in the numerous lakes; and for hiding places they hollowed out caves in the rocks by the banks of the rivers. It is more than likely that there were early British forts on the Castle Rock of Edinburgh and at Roslin, and that the caves at Hawthornden and Roslin, in the precipitous banks of the Esk, were used as places of retreat from enemies. On the western edge of Lochend, a small lake near Restalrig, there were discovered in 1871 remains of a timber framework, which is believed to have been the foundation of a crannog or lake-dwelling: and it is probable that a similar construction once existed in Duddingston Loch. Brochs —circular dry-stone edifices with very thick walls surrounding an open central court—are more frequent in the northern counties of Scotland; but vestiges of one at least remain in Midlothian. At Bow a few miles south of Stow on the north bank of the Gala, a steep hill rises 450 feet above the level of the stream, and on the summit may be traced foundations of a circular wall $13\frac{1}{2}$ feet thick surrounding a court $31\frac{3}{4}$ feet in diameter. The wall of a broch usually contained chambers at different levels with inter-communicating stairways; and a long low passage led through the wall into the unroofed inner court.

A space of three acres of the summit and eastern slope of Kaimes Hill in Ratho parish was evidently in early times enclosed by a double ditch and a rampart with an entrance on the eastern side; and on South Platt Hill there are remains of a similar place of strength. Such enclosures were probably meant to protect the rude huts of the natives and for folding their cattle during an invasion by a hostile tribe. The earthen ramparts were likely the bases into which a palisade of wooden stakes was driven. In the parish of Heriot there appear to have been several of these camps, consisting usually of three or more encircling walls and ditches, with marks of gateways, as at Borthwick Hall Hillhead and at Midhillhead.

On Heriot-town Hillhead is a circle of high stones 70 or 80 feet in diameter, and on Dewar Hill is one large stone, known as "Lot's Wife." These and similar monumental erections may have been connected with tribal religious ceremonies.

Tumuli near Mortonhall may represent an early fort; and this is rendered more probable by the occurrence in their neighbourhood of a standing stone known as the Kel Stane or Caiy Stane, a whinstone block 9 feet above ground and 4 feet below, pointing edgewise north and south. It is said to mark the site of a battle and is sometimes identified with the Camus Stone, which according to tradition gave the name to the estate of Comiston, on which it stands. Near it were found stone cists containing skeletons. Such cists exhibit one primitive method of disposing of the dead; another is exemplified in the finding during the summer of 1899, on the Braid Hills near by, of two cinerary urns now in the Antiquarian Museum. They are made of burnt clay in a flower-pot shape and are ornamented with a herring-bone pattern. They belong to the Bronze Age; and similar urns have been found at Musselburgh, Boroughmuirhead, Murrayfield, Arthur's

Seat, and Juniper Green.   Some years ago a regular
cemetery of this type was discovered in a mound at
Magdalen Bridge, east of Joppa.   The dead were first
cremated, then the ashes were gathered into the urn,

Urn found at Magdalen Bridge, Joppa.   (16 inches in height)
[*Proc. Soc. Antiquar. of Scotland*]

which was deposited in a specially prepared grave and
protected with stone slabs.

These discoveries and the finding of bronze swords
and axes on Arthur's Seat and of a small armoury of

swords on the site of Grosvenor Place, Edinburgh, point
to the inference that the district was even in those dim
ages a place of some importance, where probably the
manufacture of weapons of bronze was carried on.

Although the Roman hold upon North Britain was
much more precarious than over the southern part of the
island, traces of their occupation of the district are by no
means scanty. To judge by the numerous relics found
at Cramond and Inveresk, both were important Roman
stations. The former seems to have been linked up by
a series of forts with Antonine's Wall, which ended on
the shores of the Forth at Carriden; and Inveresk appears
to have been the objective of the main Roman road from
the south, which, leaving Hadrian's Wall near Corbridge
on the Tyne, passed north over the Cheviots and by the
great camp at Newstead near Melrose, crossed the Tweed,
traversed Lauderdale and climbed over Soutra Hill into
Lothian. Parts of this road were known later as Dere
Street and the Royal Road. There is evidence also of
a branch road skirting the east end of the Pentlands and
passing by Bowbridge and Fairmilehead to Cramond.
Numerous coins, medals, inscriptions, altars, fragments
of pavement, have been recovered in the vicinity of the
Almond mouth; and on the Hunter's Craig, a rock to
the west, the figure of a Roman eagle is cut in bold
relief.

Inveresk is understood to have been not only a military
station but also a *colonia* or civil settlement. The Praetorium probably stood on the site of the modern church,
and a road led across the Esk to the harbour near

Fisherrow. The so-called Roman bridge near Musselburgh railway station marks the line of this road, which
may have been continued to join the Cramond route by the
old road known near Portobello as the Fishwives' Causeway. In 1565 a cave with an altar to Apollo was discovered and was described in a letter sent to Sir William
Cecil by Sir Thomas Randolph, the English ambassador.
Close to Inveresk House a Roman bath was uncovered in
1783. It consisted of two rooms, the floors of which
were supported on pillars and formed the roof of the
hypocaust or lower heating chamber, where some of the
charcoal used was actually found. Urns, coins, and medals
have been unearthed at various times.

Fragments of Roman pottery and coins of Vespasian,
Hadrian, and Marcus Aurelius have been recovered at
Castle Greg on the north-western end of the Pentlands,
near Causewayend. On Wardlaw Hill and Ravelrig in
Currie parish are traces of Roman encampments. The
remains of a fort on a ridge at Longfaugh in the parish
of Crichton are supposed to be Roman; and the situation,
with its wide prospect, is certainly characteristic of the
Roman military genius for seizing important strategical
points. Near Mavisbank House on the banks of the
North Esk was a Roman station, now marked by a circular earthen mound surrounded by ramparts; and the
"Cast," a deep and narrow road at Springfield, may have
been the *via ad castra*. Wallace's Camp, so-called, a
crescent shaped fortification on Bilston Burn, may be
Roman. On Camphill (650 feet) in Newbattle parish
a quadrangular area three acres in extent was enclosed

by a rampart. It is now covered with trees. Similar
forts are to be found at Muirhouse and Craigend in Stow
parish.

Enterkin's Yett in the parish of Currie was the scene
of a traditional fierce encounter between the natives and
the Norse invaders led by Enterkin. Constantine, another
Viking leader, is said to be commemorated by the Cat
Stane, a solitary monumental stone on the peninsula
between the Almond and Gogar burn, three and a half
miles from Corstorphine.

Sculptured stones of Celtic type have been found at
Lasswade and at two places in Edinburgh, Princes Street
Gardens and Young Street. An ornament of twisted gold
wire found in Holyrood Park is one of many indications
that a love for personal decoration was not unknown
among our Celtic predecessors.

18. Architecture—(a) Ecclesiastical.

Special characteristics of national architecture depend
on the peculiar needs and customs of the people, the
climate of the region, and the nature of the materials
available; and this is true not only of the utilitarian
side of the art but still more of the decorative side,
which is the invariable complement of the other; for
decoration will be governed and guided by the natural
features and objects in the midst of which a people
lives and by the aesthetic instincts thereby produced.
Inter-communication, when it is sufficiently developed,

is apt to modify local ideas and tends to eliminate special features.

In our country the earliest kind of building construction was the erection of the rude stone monuments noticed above; but, when we speak of architecture, we usually refer to something more advanced, and we may accordingly say that the earliest style of architecture of which examples still exist in our country was the Saxon—characterised by simplicity, its masonry coarsely dressed, and the corner-stones placed alternately "long and short." To this succeeded the Norman, whose chief marks are round-headed openings, flat buttresses, and rich carving on doorways and archways.

The round-headed opening developed into the pointed arch which is the outstanding characteristic of the Gothic. This style is usually subdivided into (a) First Pointed or Early English; (b) Second Pointed or Decorated; and (c) Third Pointed or Perpendicular.

The Renaissance led to imitation of classical models in architecture as in other branches of knowledge, and the Greek pillared styles—Doric, Ionic, and Corinthian, distinguished chiefly by the treatment of the capitals or heads of the columns—exerted considerable influence on our national art. In Elizabethan architecture, which is a form of Renaissance, Classic and Gothic features are combined.

Modern Architecture is mainly imitative. Churches are usually built in some modification of Gothic, while for civic buildings the Renaissance style is most frequently adopted.

The earliest forms of ecclesiastical architecture were
the rude cells of the Celtic missionaries, one of which is
believed to have been the first predecessor of the Church
of St Cuthbert, Edinburgh. These humble edifices were
succeeded by more pretentious buildings raised under the
pious influence of Queen Margaret. The old chapel,

St Margaret's Chapel

named after her, which stands on the summit of the
Castle Rock, is the oldest ecclesiastical building in the
county: its date is not later than the time of Margaret's
sons. It is but a small room $16\frac{1}{2}$ feet by $10\frac{1}{4}$ feet within
the nave. It was restored, with the addition of stained
glass windows, in 1853.

In the twelfth century many religious orders were introduced into Scotland, and brought with them new arts
and industries. They were especially church builders and
the reign of David I, that "sair sanct for the croun," was
notable for great activity in this pious work. Many of
the Scottish abbeys owe their foundation to him—among
them Holyrood, whose institution, according to legend,
was due to the king's gratitude for the miraculous inter-
vention of the Holy Cross or Rood between him and an
infuriated hart which had turned upon him in the forest
of Drumselch. Where the sacred emblem had appeared,
the king caused the abbey to be built. In its architecture
Early English predominates, though the great east window
with its intricate tracery belongs to a later period, when
the French flamboyant influence was strong; and the
curious windows over the doorway are in the style of
the Tudor period. In spite of mingled styles the whole
effect must have been one of rich and majestic beauty.
Much of the abbey was destroyed by the English army
in 1544, but the roofless nave still remains. It was
restored by Charles II as a Chapel Royal but was again
roughly handled at the time of the Revolution by a mob
of ultra-Protestant citizens angry at its having been used
by James II for Popish worship. It was once more restored in 1758, when a roof of stone was put on, but ten
years later this fell in. Since then, except for preservation
work, it has been left a stately ruin, a recent proposal to
rebuild it having been vetoed by King Edward VII. It
contains the tombs of several royal personages and other
notabilities. The Abbey of Holyrood was occupied by

Holyrood Chapel

monks of the order of St Augustine, whose mills gave name to a district of the city—Canonmills—and who under royal grant erected a burgh of barony between the city gates and the abbey—Canongate, which remained a separate burgh until quite recent times. The abbey had also the privilege of sanctuary; and before imprisonment for debt was abolished, its precincts were much resorted to by debtors.

The numerous churches erected in the twelfth and thirteenth centuries were built in the styles then prevailing, the Norman and the First Pointed Gothic. Though they have suffered severely at the hands of invader, reformer, and restorer, their number and magnificence give evidence of the prosperity of the country in those days. But the stormy times of the Wars of Independence succeeded, when Scotsmen were more familiar with the clash of arms than the sound of hammer and when the "fair abbayes" suffered frequent damage from the invading armies. Thus the fourteenth century is devoid of examples of church architecture. When in the fifteenth century church building was revived it was in a country impoverished by the long wars, and the edifices are as a rule of a less pretentious type than the great abbeys of earlier and more prosperous days. Barons built and endowed chapels on their own estates or in neighbouring towns. The endowment usually specified the number of clergy attached to the church and known as the "College." The most famous of these Collegiate churches was that founded by Mary of Gueldres, the queen of James II— Trinity College, which stood till 1848 on part of the

ground now occupied by the Waverley Station. It consisted of choir, transepts, and central tower; and the
design and execution were as fine as in the best Gothic

St Giles', Edinburgh

structures in England. Many of its features were reproduced in the modern Trinity College Church in Jeffrey
Street.

Roslin Chapel

The old church of St Giles', which had been burnt
by the English in 1385, was restored and made Collegiate
in 1466, since when at different times it has been enlarged
and altered. The general plan of the church is cruciform,
the lantern tower, "the ancient croon o' stane," springing
from the meeting of the arms. Aisles lead out of the
main transept. The roof is groined, and the windows
are all in the Perpendicular Gothic style, filled in with
fine stained glass. At the Reformation the church was
divided into four compartments, three of which were
used as places of worship. Round the walls booths of
merchants were attached and these were known as the
"Krames." Largely by the munificence of Sir William
Chambers, the publisher, St Giles' was restored to its
original unity. It is now regarded as the national church
of Scotland and the depository of the shot-torn flags of
Scottish regiments, and of other memorials. Through a
bequest of the Earl of Leven and Melville a beautiful
chapel for the Knights of the Thistle has lately been
added at the south-east corner of the main building.
The most outstanding incident in the history of St Giles'
is the riot which took place when the New Liturgy was
read, 1637.

Roslin Chapel on the north bank of the Esk is also
one of these fifteenth century Collegiate churches, and is
a fine example of the Decorated or Florid style of Gothic.
In spite of cramped outline and want of symmetry, it is
peculiarly interesting and attractive on account of the
variety and elegance of detail in its window tracery, its
thirteen different kinds of arches, its flying buttresses,

Roslin Chapel and Prentice Pillar
(with spiral carving)

and its pinnacles and mouldings. Concerning one of its pillars, the Prentice or Prince's Pillar, the familiar legend is related of the apprentice of genius who in his master's absence executed the work in so fine a manner as to incur the jealousy of the master, who killed him by a blow of his own hammer. The chapel was founded in 1446 by

Corstorphine Church

William St Clair, whose family had a hereditary connection with the Masonic craft. Only the chancel was completed.

Corstorphine Church, recently restored, was founded in 1429 by Sir John Forrester; and the monuments to the Forrester family are specially interesting. The church is in the form of a cross with an additional transept on one of the sides.   The style is mainly decorated Gothic.

On the site of an earlier building, which was at one time included in the diocese of Dunkeld, stands Cramond

Crichton Church

Church, its oldest portions dating from 1656.   The building is cruciform in shape and has a square embattled tower.   In the churchyard are many curious old tombstones, including that of the Howiesons of Braehead, whose founder is said to have rendered timely aid to James V in one of his escapades as the Gudeman of Ballengeich.

In pre-Reformation times Restalrig Church was famous as the shrine of St Triduana, who died at Restalrig and who was specially helpful to persons suffering from eye-diseases. A Collegiate church was established there by James III in 1487; but the first General Assembly ordained that the church "as a monument of idolatrie be raysit and utterlie casten downe and destroyed." An octagonal chapter-house with groined roof, adjoining the church, has lately been restored.

Other county churches worthy of notice are those of Crichton, Dalkeith, Torphichen, Midcalder, and Duddingston. At the gateway of the last named still hang the *iougs*, an iron collar and chain by which offenders were confined and exposed to the censure of their pious neighbours; and close by is the *loupin'-on stane*, where ladies mounted their horses on leaving church.

Of the Edinburgh churches built in recent times, we can mention only St Mary's Cathedral. This magnificent fabric with its spire rising to almost 300 feet is one of the most important churches reared in Britain since the fifteenth century. Its general style is Early English; and it was planned by Sir Gilbert Scott. The foundation stone was laid in 1874, and the cathedral was formally opened in 1879.

The Jougs, Duddingston Church

19. Architecture—(*b*) **Military.**

It has been already noted that in Scotland as elsewhere the earliest form of fortification was probably the earthen mound surrounded by a wooden palisade on a turf wall. The crannog and the broch were common, the latter specially in the north. Building in stone and lime was perhaps introduced by the Normans who flocked into the country in the eleventh and twelfth centuries. Although no original Norman keeps remain, their style was imitated in the peel towers which were built in many places after the War of Independence. Previous to the war the thirteenth century was a time of prosperity in Scotland and many castles were built after the older model—a large fortified enclosure, usually quadrangular in shape, with the angles strengthened by towers and the gateway defended by a portcullis. But the exhaustion of the war changed the character of these places of defence. Simplicity and strength were the essentials, and the fashion of the Norman keep was adopted.

The simple keep consisted of a rectangular tower with thick walls ending at the roof in a projecting parapet, from which the fort was defended and which had therefore openings or machicolations to permit of missiles being launched upon assailants. The ground floor was vaulted and was chiefly used as a store-room and a place of protection for the baron's cattle: it had also a loft, where the menials slept. The entrance to the tower was on the second story and was reached by a removable ladder.

Here was the main apartment of the tower—the hall, also vaulted and frequently divided into an upper and a lower room by a wooden floor resting on stone corbels halfway up the walls.   The third story was a repetition of the second, and the vaulted ceiling carried the stone

Dalhousie Castle

roof of the tower.   This story contained the private apartments of the lord, which were reached by an internal stair leading also to the parapet.   A courtyard surrounded by a strong wall was usually attached to the keep as a first means of defence.

Liberton Tower, two miles south of Edinburgh, was

a simple keep of this kind, though it was built in the fifteenth century and is only two stories in height. So also are Craiglockhart and Cakemuir, the latter famous

Merchiston Castle

as having sheltered Queen Mary when she fled in male attire from Borthwick Castle, three miles away, to meet Bothwell.

The simple keep form afforded little accommodation

of a private kind, and this want caused the various modifications known as the L, the Z, the E, and the T plans, and also an approximation to the earlier castle form of an encircling wall with strengthening towers. Dalhousie Castle, now much altered and enlarged, was originally a keep with one wing on the L plan. Such also were Merchiston Castle, Hatton House, Inch House, and Bavelaw Castle. Borthwick Castle was a keep with two wings and thus approached the E plan. The Z plan is exemplified in Craigcrook Castle, Ravelston House, and Riccarton House. It was adopted when firearms became general, the two wings enabling the defenders to rake the approach to the main building on all sides. Baberton House is built on the E plan.

The courtyard gradually developed in importance, and along the inner face of the wall of *enceinte* a number of outhouses came to be built to provide the further accommodation required. Thus Craigmillar and Crichton are keeps which have been extended into castles by the addition of buildings round the courtyard. Indeed Edinburgh Castle and Holyrood Palace may be regarded as of this type, which, as times became more settled and defence could be subordinated to comfort, developed more and more the character of a mansion built round a quadrangle. But the mansion continued to retain many of the distinctive features of the fort, and thus the Scottish Baronial style of architecture, with its crowstepped gables, its corbellings, and its turrets, was evolved and gave to the buildings of the seventeenth century and later times their peculiar picturesqueness.

The castle of Edinburgh has been so frequently altered that nothing of earlier date than the fifteenth century remains, with the exception of St Margaret's Chapel already noticed. The old Parliament Hall in the upper part of the castle was long used as a military hospital but has now been restored to something like its former dignity. The Royal Palace near by was used as a residence by kings

Scottish Regalia

and queens of Scotland for several generations. The most interesting room in it is that in which Queen Mary gave birth to James VI in 1566. Its old ceiling of panelled wood with painted monograms and crowns still remains. The Crown Room contains the ancient regalia, "the Honours of Scotland," deposited there in 1707 after many vicissitudes. Mons Meg, the oldest cannon in

Europe, occupies a position of honour on the King's Bastion, the highest point of the rock, which commands a magnificent view of the city, the Firth of Forth, and the Highland Hills.

Craigmillar Castle stands on the edge of a cliff which drops 20 feet to the slope beneath; and on this side, the

Mons Meg

south, the original entrance was placed. The great hall has a vaulted ceiling, a fine pillared fireplace, and window recesses in the 9 feet thick wall. A small room off the hall is known as Queen Mary's Room, the castle having at various times been occupied by her. An old sycamore tree at the foot of the castle hill is said to have been planted by her, and the village close beside it, Little

M. M.   9

France, got its name from being the residence of her French attendants. At an earlier date, 1477, Craigmillar was the scene of the imprisonment of the Earl of Mar, brother of James III. In 1660 the castle passed from the possession of the family of Preston, who held it in Queen Mary's time, to that of Gilmour, whose present representative keeps the building in an excellent state of repair.

A rocky promontory almost surrounded by a loop of the North Esk affords a good vantage ground for the castle of Roslin. The oldest part now remaining, a tower near the entrance, dates from about the beginning of the fourteenth century. Additions at later times made the castle a place of considerable size and importance. One of its owners, St Clair, third Earl of Orkney, kept a state almost kingly. In 1544 the castle was among those destroyed during Hertford's invasion ; and after being restored it was battered and captured by Monk in 1650. The caves beneath the castle were used partly as rooms for servants and partly as dungeons. Scott's ballad of *Rosabelle* celebrates the castle and its lordly owners. Of an older building, the Maiden Castle, which stood near by in another link of the Esk, not even the ruins are left.

Borthwick Castle occupies a strong natural position on a tongue of land between two head waters of the Gore, the South and the North Middleton Burn, to whose banks the castle ridge runs steeply down on three sides. The place was originally known as the Mote of Locherwort and took its later name from its founder, Sir William Borthwick, whose earlier seat was at Catcune

Borthwick Castle and Church

Castle a mile and a half further down the Gore. The walls of the keep are 14 feet thick at the bottom and 6 feet at the top, and rise to a height of 110 feet, where they sustain a roof of stone slabs. The entrance to the great hall, a magnificent apartment, was by an outer stair and drawbridge; and the castle was further defended by a strong wall round the courtyard. It is chiefly famous

Crichton Castle

for its association with Queen Mary and Bothwell in 1567, and for its stout resistance in 1650 to Cromwell, who had to "bend his cannon against it" before the garrison under the tenth Lord Borthwick capitulated. The unusual height to which the walls were carried is probably explained by the fact that this made possible communication, by signalling, with the castle of Crichton a mile and a quarter to the east. The position of both castles was doubtless chosen with a view to commanding the road from the south.

Crichton Castle "rises on the steep of the green vale of Tyne," and its original form was a square tower of the earliest type of baronial architecture dating back to the fourteenth century; but at a later time this keep was extended into a castle surrounding a courtyard, the additions exhibiting great variety and lightness in details. Scott brings Marmion to the castle on his way to Edinburgh, and speaks of it thus :

> " Nor wholly yet had time defaced
>   Thy lordly gallery fair ;
> Nor yet the stony cord unbraced,
> Whose twisted knots with roses laced
>   Adorn thy ruined stair.
> Still rises unimpaired below
> The courtyard's graceful portico ;
> Above its cornice, row on row
> Of fair-hewn facets richly show
> Their pointed diamond form,
> Though there but houseless cattle go
> To shield them from the storm."

In the reign of Queen Mary, Crichton belonged to the Earl of Bothwell, and the queen was present there at the marriage feast of Sir John Stewart.

20. Architecture—(c) Municipal.

The most ancient municipal building was the Tolbooth, which in origin was the strong house of the civic authority in the burgh, and some part of which invariably served as the town prison. It usually had a tower with a bell to mark the time or to summon the burghers to arms. At a later date another use became more prominent as the burgh grew in size and importance—the Tolbooth came to be the Council Chamber. The prison was by and by disjoined, and this permitted the building to be made more elaborate.

Thus the Canongate Tolbooth, erected in 1591, has a tower representing the earlier keep type and serving as belfry and prison. Adjoining the tower and entering by an outside stair is the Council Chamber of later date.

The Edinburgh Tolbooth, "the Heart of Midlothian," one of the scenes of the Porteous Riot (1736), stood in the middle of the High Street, but was removed in 1817. Its site is now marked by a heart-shaped arrangement of causeway stones in the footway between the Buccleuch Statue and St Giles' Cathedral.

Musselburgh Tolbooth consists of three stories, each vaulted. It is of massive construction with bartizans and parapets projected on corbels. The entrance to the Council Chamber is on the east by an outside stair. An older Tolbooth was destroyed in 1544 and the present was erected in 1590, but part of the tower may belong to the original edifice. The materials used in the reconstruction were taken from the ancient chapel of Loretto and this sacrilege drew upon the town the annual sentence of excommunication passed by the Pope for about two

Town Hall, Musselburgh

hundred years.   The Renaissance Council Chamber was added in 1762.   Thus the three stages are all represented —the keep, a strong sixteenth century structure, and a late council chamber.

Old Parliament House, Edinburgh

Included in the range of buildings which form Parliament Square behind St Giles' Cathedral, Parliament House is most impressive in an interior view, when its great size and simple grandeur are appreciated,   The roof formed of dark oaken beams adjusted to the outline of a circular arch with numerous gilt pendants is specially fine.   The

George Heriot's Hospital

Hall is built on the old churchyard of St Giles' and dates from about 1639. It was the scene of the stormy debates preceding the Union of the Parliaments in 1707; was thereafter divided up into booths; but is now cleared and used as a promenade for advocates, solicitors, and clients, being "the Hall of Lost Footsteps."

Many of the educational buildings are worthy of note. George Heriot's Hospital may be described as full of architectural contradictions. From a distance its turrets, pinnacles, and chimneys appear to be grouped without design; nearer, one seems to find symmetry of disposition between the halves of the building; and on a still closer inspection the surprising truth is found to be that, except in one case, no two parts are exactly alike, Gothic, Roman, and Corinthian styles being curiously mingled. The Hospital is built about an interior quadrangle with four square towers at the corners, the towers themselves having turrets at their angles. The connecting curtain wall is broken in front by the segment of an octagonal tower with Gothic windows. The interior basement story is arcaded round the paved quadrangle, and a statue of the pious founder, "Jinglin' Geordie" (see Scott's *Fortunes of Nigel*), is placed in a niche over the vaulted entrance. Inigo Jones is doubtfully credited with the main design; the execution appears to have been carried out by various master masons, notably William Wallace of Tranent. The building, begun in 1628 and finished in 1650, was occupied by Cromwell as a military hospital after the Battle of Dunbar. Originally an institution where the foundationers were boarded and educated, it is now an important secondary school, specially equipped on the scientific and technical sides.

A similar institution, which however has retained its original purpose, is Donaldson's Hospital, where boys and girls—especially, deaf mutes—are boarded and taught. The hospital occupies a splendid site in the western

Donaldson's Hospital

suburb of Murrayfield above the gorge of Leith Water. It was built (1842–51) from designs by W. H. Playfair and is a fine palatial structure in the Tudor style, arranged round a quadrangle.

The General Post Office occupies a site where stood formerly the Theatre Royal and Shakespeare Square. It is in Italian style and has several times been added to since

Dean Bridge, Edinburgh

it was built in 1861–66. Opposite it is the Register House, where the national records are kept, a handsome building also Italian in style with fine Corinthian decoration. An equestrian statue of the Duke of Wellington stands in front of the entrance.

The alternation of ridge and hollow in the contour of the site of the city makes Edinburgh a place of bridges. Of these the North Bridge, joining the New Town and the Old, and the Dean Bridge, over the gorge of the Water of Leith, are the most notable. The latter was built after designs by Telford and crosses the valley by four fine arches at a maximum height of 106 feet from the bed of the stream. The view eastward from the bridge is remarkably picturesque. A prominent feature in the valley is the Greek portico, with statue of Hygeia, built over St Bernard's mineral well, the properties of which are not unlike those of more famous spas.

21. Architecture—(*d*) Domestic.

As in other places the houses of ancient Edinburgh were mere rude wooden and thatched huts huddling under the protection of the castle; but by the beginning of the sixteenth century these town houses began to be constructed of masonry. To the front, however, a wooden framework jutted out forming a covered arcade on the street level and an open verandah for the upper floors. On to the arcade opened the booths of the merchants; and the galleries above gave opportunities for the tenants to take the air and study the passers by. The closes, which led down the slopes of the ridge from the main street, were sometimes common passages and sometimes private entrances to courtyards round which the town house of a noble or of a prosperous merchant was built. When space became more valuable the wooden galleries were boarded up and gave additional rooms to the house, which had thus a false front of timber and the real stone one some seven or eight feet within. Later this type of house was built with stone fronts, the arcade still being preserved and the upper floors in some cases projecting on corbels each a little beyond the one beneath, as was the case with an old house which stood till near the end of last century at the corner of the Lawnmarket and the West Bow. Usually an outside stone stair led up to the first floor and further communication was by internal stairs of wood. The so-called "Speaking House" in the Canongate, with its quaint mottoes, is now the only specimen left of the old timber-fronted house once so common.

Within, these houses were frequently richly decorated, the walls in panelled wood, the ceilings in ornamental plaster work of quaint and beautiful design. As the influence of the Renaissance spread the architecture of the town house lost its peculiar Scottish characteristics.

John Knox's House stands at the foot of the High Street close to where the Netherbow Port once formed the eastern entrance to the town. Though the house is now divided into various portions, it was originally one large house of four stories with a "laigh" floor under the street level and also attics. The building is peculiar from the fact that the projections superadded to the main block are partly of stone and partly of timber. Some of the details of the stonework are interesting, notably the figure of Moses carved above two sun-dials, and pointing to a sun inscribed "ΘΕΟΣ, Deus, God"; and also the motto on the frieze over the street floor windows :

LVFE · GOD · ABVFE · AL · AND · ɎI · NYGHBOUR · AS · YI · SELF.

John Knox's House

The connection of John Knox with the house is hardly sufficiently authenticated.

Bailie Macmorran's House in Riddle's Close is an excellent example of the courtyard house as adapted to town conditions and used partly for residence and partly for business purposes. The rich internal decorations were striking evidence of the wealth and importance of the owner—the prominent merchant and magistrate whose tragic death in the "barring-out" escapade of the High School boys in 1595 is related in Chambers's *Domestic Annals*.

Of all the residential buildings the most interesting historically is Holyrood Palace. The abbey, which has already been noticed, was frequently resorted to by Scottish sovereigns from Robert Bruce to James IV, the latter of whom founded close beside it a royal palace; and this was enlarged by his successor, mostly destroyed by Hertford in 1544, rebuilt immediately thereafter, accidentally burnt while occupied by Cromwell's troops, and finally reconstructed by Charles II. The plan includes two square towers, with round towers at their front angles, these angular towers terminating in conical roofs in Scottish Baronial sixteenth century style. Between the two main towers runs a lower screen wall recessed from the front line of the towers and pierced in the middle by the main entrance, which is surmounted by a crown. The south tower and the greater part of the connecting masonry of the quadrangle are late in date, but the north tower is part of the sixteenth century building and contains the most interesting part of the

Holyrood Palace and Abbey

palace, Queen Mary's rooms, which are maintained as when she lived in them. The Picture Gallery, with its rows of wretched portraits, mostly imaginary, of Scottish kings, many of them fictitious, was the scene of Prince Charles Edward's receptions in 1745. It is now used as the meeting-place of the Scottish Peers when they assemble at the beginning of each Parliament to elect sixteen

Queen Mary's Bedroom, Holyrood

representatives to the House of Lords, and as the levee-room of the Lord High Commissioner, during the annual sittings of the General Assembly of the Church of Scotland.

The New Town exhibits some fine examples of domestic architecture, especially in the squares, crescents, circuses, and terraces, which were laid out in the latter part of the eighteenth and the beginning of the nineteenth century.

It was noted above that the courtyard type of castle developed into the mansion as defence became less necessary. At the same time the interiors also underwent a transformation as comfort came to be studied, and increased

Dalkeith House

wealth employed art and elegance to decorate and beautify the home. Of the numerous splendid mansion houses in Midlothian only a few can be noticed.

Arniston, an eighteenth century house of massive and

imposing appearance, is finely situated in extensive grounds through which flows the Esk.

Carberry Tower occupies a commanding position on the northern slope of Carberry Hill.

Dalhousie Castle is a fine old baronial mansion pleasantly situated on the banks of the South Esk.   The

Hawthornden Castle

strong defences of the ancient castle have been modified to adapt the house to modern requirements.   Cockpen House which once stood on the opposite side of the river was bought by the Earl of Dalhousie, who is accordingly the titular "Laird o' Cockpen."

Dalkeith House or Palace stands on the site of a much older castle, on rising ground overlooking the North Esk near the town of Dalkeith. It was built in 1705 by Anna, Duchess of Buccleuch, in imitation of the Palace of Loo in the Netherlands. The house contains a fine collection of paintings, and the grounds extend to a thousand acres in a landscape of much natural beauty. Margaret Tudor, Cardinal Beaton, Regent Morton, James VI, Charles I, General Monk, George IV, Queen Victoria, and Edward VII are all associated with the house or its predecessor.

A specially fine example of a seventeenth century Scoto-French mansion ingeniously added to a tower of the fifteenth is Hatton House, whose combination of quaint windows and tall chimneys gives it a highly picturesque appearance.

Hawthornden, romantically situated on the bank of the North Esk, was the property and residence of William Drummond the poet, who here entertained Ben Jonson, as the distich tells :

"Welcome, welcome, Royal Ben!"
"Thank ye, thank ye, Hawthornden!"

In the cliffs underneath the house are several artificial caves, which were probably used as places of concealment in troublous times.

Near the village of Lasswade on the north bank of the Esk in magnificently wooded grounds stands Melville Castle, built in 1786 on the site of an older house which belonged to Melville, Queen Mary's secretary.

Pinkie House, Musselburgh

One of the most interesting of Midlothian mansions is Pinkie House near Musselburgh. The building is of different periods, the older part being a massive square tower with picturesque turrets at its angles. In 1613 the house was greatly extended by its owner, the Earl of Dunfermline, who added its most interesting room, the Painted Gallery, the roof of which is lined with wood

Oxenfoord Castle

painted with coarse but powerful representations of heraldic and mythological subjects. Prince Charles Edward is said to have stayed in the house the night after the battle of Prestonpans.

Craigcrook, once the residence of Lord Jeffrey; Merchiston, formerly the seat of the Napiers, especially him of logarithmic fame—the auld wizard; Newhailes,

built by the famous lawyer and antiquary, Sir David
Dalrymple, Lord Hailes; Newhall, birthplace of Dr
Alexander Pennicuik, poet and physician; Niddrie-Mari-
schall, home for many generations of the Wauchopes, and
beloved by Lord Cockburn; Oxenfoord, a magnificent
edifice in the valley of the Tyne, the seat of the Earl
of Stair; Whitehill, a fine example of Jacobean style;
Woodhouselee, successor to the "haunted Woodhouselee"
of Scott's *Grey Brother*; all of these are worthy of note
did space permit.

22. Communications—Past and Present.

Like those in other parts of Britain previous to the
middle of the eighteenth century, the roads of Lowland
Scotland were little more than beaten tracks winding along
hillside and through valley. "Made" roads were practically
unknown until the passing of the first Scottish Turnpike
Act in 1750. Two great roads long survived in Southern
Scotland to testify to the energy and engineering skill of
the Romans; but during the Middle Ages road-making
was a lost art, and in the beginning of the eighteenth
century the condition of the roads may be inferred from
the fact that the contractor who undertook to run coaches
between Edinburgh and Glasgow did not guarantee to
complete the double journey in less than six days. Sixty-
five minutes now suffices for an express train to convey
its passengers from the one city to the other. By 1773
matters must have considerably improved, for Dr Samuel

Johnson travelled to Edinburgh from London not by sea but in post-chaises, " of which the rapid motion was one of his favourite amusements."

The main lines of communication from Edinburgh are determined by the physical configuration of the county ; so that we have routes (1) to the east and the west along the coast ; and (2) inland to the south following the river valleys and skirting or crossing the hill ridges of the Pentlands, the Moorfoots, and the Lammermoors. Thus we have the coast road to the east and south *viâ* Portobello and Musselburgh, leading by Prestonpans, North Berwick, and Dunbar to Berwick and thence to Newcastle and London. This road divides at Levenhall near the county boundary, one branch keeping close to the sea and one striking inland, climbing to the top of the 100 feet raised beach, and passing by Tranent, Haddington, and East Linton to Dunbar.

Westward, the coast road is known as the Queensferry Road, leading as it does to the ancient crossing place over the Forth, named after Malcolm Canmore's sainted queen. This road was long the main route to the north-east, and is immortalised in Scott's *Antiquary* as that traversed by the " Hawes Fly " or Queensferry Diligence—" green picked oot wi' red ; three yellow wheels an' a black ane."

To England two other routes may be followed. The first leaves the city by Newington and Little France, passes by Dalkeith and over Soutra Hill (1209 ft.)—where once stood the church and hospital of the Holy Trinity, which was founded for pilgrims about 1164 and of which only a small aisle remains—and continues by Greenlaw and Coldstream. The second goes by Gilmerton, Eskbank, and Soutra to Lauder, on by Earlston to Jedburgh, crossing Carter Fell (1350 ft.) and so to

The Luggie, Soutra

Newcastle. A third route, of old dominated by the castles of Borthwick and Crichton, follows the valleys of the Esk, the Gore, and the Gala to Galashiels and Hawick.

The Esk valley is also used as far as Penicuik by the Peebles road, which passes thence by Leadburn to the valley of Eddleston Water, and so into Tweeddale, and from there by Tweedsmuir over the Lowthers and the Devil's Beef Tub to Moffat.

Clydesdale is reached by either of the roads running on the north and south sides of the Pentlands, the former by Currie, the Lang Whang, and Carnwath to Lanark, and the latter by West Linton and Biggar. Lanark may also be reached *viâ* the three Calders, Wilsontown, and Forth.

To Glasgow the old road was by Midcalder and Shotts ; another route traverses the busy mining centres of Broxburn, Uphall, Bathgate, Armadale, Airdrie, and Coatbridge ; and a third goes by Linlithgow and Falkirk. This last leads also to Stirling and the north.

Besides these main routes cross-roads innumerable cover the county in all directions.

Road and rail alike follow the easiest possible gradients, and it is not surprising therefore to find that their tracks are often parallel. The well-known East Coast route to England runs alongside the coast road by Dunbar and Berwick. This was the original line of the North British Company, which was formed to work the railway between Berwick and Edinburgh. The Waverley route takes advantage of the Esk, Gore, and Gala valleys, the ridge between the two last being crossed at a height of 900 feet near Heriot station. The line continues by Hawick to Carlisle. A third route to England is that by the Caledonian Railway, which passes west into

Lanarkshire, joins the Glasgow line at Carstairs, and follows Clydesdale and Annandale to Carlisle.

Three lines connect with Glasgow: (1) the North British by Linlithgow and Polmont; (2) the North British by Bathgate and Airdrie; and (3) the Caledonian by Midcalder and Shotts.

Both these companies provide communication also with the north. At one time the North British traffic with Fife, Dundee, and Aberdeen was conducted by ferries over the Forth between Granton and Burntisland or between South and North Queensferry, and over the Tay between Tayport and Broughty Ferry; but the erection of the Forth and Tay bridges has greatly improved these parts of their service. The Caledonian route makes use of the North British line to Larbert and crosses the Forth at Stirling.

Short lines from Edinburgh connect with Leith, Granton, and Corstorphine; and a circular railway, the Suburban, serves the outskirts of the city. There are besides, branch lines from the main lines to numerous centres of population in the county.

Tramways worked by underground cables serve the city and connect at Joppa with an electric tramway, which passes on by Musselburgh and Prestonpans to Port Seton, following the shore road. Leith tramway system is also electric; and a short line has recently been opened from Dalry to Slateford, whence extensions are under construction to Redford near Colinton, where new cavalry and infantry barracks are being built.

The first tramways in Scotland were built to transport coal; and it is interesting to note that the first to carry passengers was one constructed between St Leonard's, Edinburgh and Dalkeith, with a branch to Musselburgh. The vehicles were drawn by horses, and on that account and perhaps also from its leisurely methods this tramway came to be known as the "Innocent Railway." Part of the line still exists, and is used as a mineral branch by the North British Railway.

At one time, too, fast passenger boats plied on the Union Canal between Edinburgh and Glasgow. The canal begins at Port Hopetoun in Edinburgh and leads $31\frac{1}{2}$ miles west through Midlothian, West Lothian, and Stirlingshire, to its junction with the Forth and Clyde Canal near Falkirk. It has been acquired by the North British Railway, and now only a little heavy barge traffic disturbs its waters.

23. Administration.

Local administration in Scotland is confusing. The system has grown up through a succession of changes and modifications, in most cases introduced with a particular end in view and with little reference to other departments of the public service. There are consequently no definite units of administration by the grouping of which larger divisions might be obtained organically connected in every respect with the smaller. To begin with, we have civil parishes and ecclesiastical parishes ; and in many cases the boundaries of these have no relation the one to the other. If a new ecclesiastical parish is set up, the civil rights of the landowner in the new parish remain related

to the older civil parish.   Indeed, until the work of the
Boundary Commissioners under the Local Government
(Scotland) Act 1889 took effect, it was no uncommon
thing to find a parish divided between two counties.
A registration county need not coincide with the civil
county ; and a parliamentary constituency may include
districts which for other purposes are in different administrative areas.

A shire or county was originally a division of the
country supervised by a royal representative, the sheriff.
Sheriffdoms in Scotland were in existence at least as early
as the reign of David I, but their number and the extent
of their jurisdiction naturally underwent frequent modification in unsettled times.   The sheriffdom of Lothian is
known to have been one of the first to be constituted ;
but, although its nominal extent was probably over all
the three modern counties of Linlithgow, Edinburgh, and
Haddington, the authority of the sheriff came to be greatly
restricted by royal grant of special jurisdictions to various
personages and communities within these bounds.   Thus
the Constable of Edinburgh Castle, the Justiciar of
Lothian, the Provost of Edinburgh, the Abbot of Holyrood, the Monks of Dunfermline, the Archbishop of
St Andrews, the Stewarts, the Douglases, and the Barons
of Duddingston, Prestonhall, and Carrington, all exercised
jurisdiction of some sort within the territory of the Sheriff
of Lothian.   All of these powers, together with the
hereditary character of the sheriffship itself, were abolished
in 1747.

The Militia Act of 1782 set up in each county

a kind of military organisation, headed by the Lord-Lieutenant, who is chosen by the Crown from the local gentry and is supported by a number of deputy-lieutenants. These positions, however, are largely ceremonial; and the chief county official is the paid sheriff, who with the help of sheriff-substitutes is responsible for many important administrative and judicial duties. The district of the sheriff of the Lothians now includes Midlothian, Linlithgowshire, Haddingtonshire, and Peeblesshire; and he is assisted by four sheriff-substitutes.

A large amount of unpaid public work is done by Justices of the Peace, local gentlemen appointed for such duties by the Lord Chancellor.

Three types of burghs are represented in the county. Edinburgh is a royal burgh—one of 55 such in Scotland. As the name indicates, these are governed by corporations which have been constituted by Royal Charter. They have also the privilege of sending representatives to Parliament apart from the county constituency within which they are situated. It is to be noted that at one time Canongate and Portsburgh were separate burghs—burghs of regality, but they were amalgamated with the city in 1856. Portobello, since 1896, has also been municipally incorporated in the city, although for Parliamentary purposes it is still conjoined with Leith and Musselburgh. These two are not royal burghs, having no Royal Charter; but the privilege of Parliamentary representation was conferred upon them, along with twelve other Scottish towns, by the Reform Acts; and they are therefore called Parliamentary Burghs.

Places with a population of more than 700 may become incorporated by adopting the Police Acts, and this has been done by Bonnyrigg, Dalkeith, Lasswade, Loanhead, and Penicuik. They are known as Police Burghs, and have powers of self-government in such matters as cleaning, lighting, paving, and generally improving the burgh, according to the extent in which they have adopted the general Acts.

The Town Council is the ruling authority in the affairs of the burgh. The councillors are elected by the householders for a term of three years, one-third being elected yearly; and the larger burghs are divided into wards for election purposes. The magistrates are chosen annually by the Council, and perform certain judicial functions in the police and licensing courts.

Previous to the passing of the Local Government (Scotland) Act 1899, county affairs were in the hands of several bodies, viz. the Commissioners of Supply, the County Road Trustees, the Local Authorities constituted to administer special Acts relating to public health, etc., and the Justices of the Peace. With one or two trifling exceptions the administrative powers of all these bodies were by the Act cited above transferred to County Councils. Roads and bridges, public health, diseases of animals, valuation, finance, and general administration are under the control of the Council. Police administration is managed by a joint committee representative of the County Council and the Commissioners of Supply. The latter are practically the land owners in the county, and their public functions as a body are now restricted to appointing their representatives to the above Committee. Midlothian and Peeblesshire together form a special district for lunacy administration, which is directed by a joint board representative of the two County Councils and of the burghs of Musselburgh and Peebles. The County Licensing Court is composed of seven elected representatives of the County Council and seven representatives elected by the Justices of the Peace.

Poor Law administration was formerly managed by Parochial Boards, consisting in each parish of so many members nominated by the Kirk Session and so many elected by the ratepayers; but the Local Government (Scotland) Act 1894 set up in every poor-law parish (or combination of parishes) a Parish Council, elected entirely by the ratepayers, which, besides taking the place of the Parochial Board, has certain other important duties entrusted to it, such as levying poor, public health, and education rates; registration; vaccination; protection of children; maintaining rights of way; managing parish trusts; and providing burial grounds.

The following is a list of the 32 parishes in the county : Colinton, Corstorphine, Cramond, Liberton, Newton, Currie, Ratho, Kirknewton, Midcalder, West Calder, Carrington, Cockpen, Dalkeith, Glencorse, Inveresk, Lasswade, Newbattle, Penicuik, Borthwick, Cranston, Crichton, Heriot, Stow, Temple, Duddingston, Fala and Soutra, City of Edinburgh, St Cuthbert's, Canongate, North Leith, and South Leith. Kirkliston is partly in the county of Edinburgh and partly in Linlithgow county.

For ecclesiastical affairs the parishes are mostly included in the Presbyteries of Edinburgh and Dalkeith, which again are embraced in the Synod of Lothian and Tweeddale. Kirkliston, however, is in the Presbytery of Linlithgow, and Stow is in the Presbytery of Earlston and the Synod of the Merse and Teviotdale.

As regards Parliamentary representation there are four constituencies in the city ; Leith, Portobello, and Mussel-burgh are combined as the Leith Burghs ; and the rest of the county forms one constituency ; so that six members are returned from the whole area.

Elementary education is controlled mainly by School Boards, first constituted by the Education Act of 1872, which set up in each burgh and parish a Board with power to impose a rate for the support of the necessary schools, the levying of the rate however being in the hands of the Parish Council. The schools are maintained chiefly by these rates and by grants from Government administered by the Scotch Education Department, which is the supreme authority under Parliament. Certain church schools are not under the control of the School Boards. Attendance at school is compulsory between the ages of five and fourteen, and in Board Schools as a rule no fees are charged for this stage of education. After the elementary stage follows what is now technically called the Intermediate course, providing in specified schools a three years' curriculum in languages, mathematics, science, and other recognised subjects. Secondary schools offer a similar curriculum of five years' duration, leading up to the attainment of the Leaving

Certificate, which is regarded as a passport to a university course.

Many of these secondary schools possess large endowments left by their pious founders. Fettes College, a conspicuous Gothic building on the northern outskirts of the city, was erected in 1865–70 and endowed under a bequest of Sir William Fettes of Comely Bank (*see* p. 191). It is conducted on the lines of the great English public schools, as are also Merchiston—the main building of which is the ancient Castle of the Napiers—and Loretto, at Musselburgh, both of which are proprietary establishments. The Edinburgh Academy in Henderson Row was instituted in 1824 by a committee of citizens among whom were Sir Walter Scott, Henry Cockburn, and Henry Mackenzie. It is a day-school and is managed by a body of directors under a royal charter. The Merchant Company Education Board has under its care George Watson's Boys' College, Daniel Stewart's College, the Edinburgh Ladies' College, and George Watson's Ladies' College : George Heriot's school is under the Heriot Trust : and the School Board controls the Royal High School and the Higher Grade Schools of Boroughmuir, Broughton, and Portobello.

In all the School Board areas throughout the county provision is made for completing elementary education in continuation classes ; and in many cases these classes offer also instruction of a specialised type relating to the main industries of the locality.

Besides the general system of education here outlined, there are several technical institutions, in which training for particular arts and crafts is obtainable, such as the Heriot-Watt College, the College of Art, the College of Agriculture, the Veterinary College, the School of Domestic Economy, and the Training Colleges for Teachers.

24. The Roll of Honour.

From early times Edinburgh and its neighbourhood have been intimately associated with royalty. After the death of Malcolm Canmore its castle afforded a refuge to his widow and family. The sainted queen is commemorated by a chapel dedicated to her name; and her sons and successors frequently resided in the Castle, and in the abbey of Holyrood founded by David I. In the times of the Stuart dynasty the town became the chief burgh of the country and was closely identified with the fortunes of that royal house. In James II's reign it was definitely recognised as the metropolis, and James IV built Holyrood Palace. Much of the story of Queen Mary has its scene laid in the city and its vicinity, various castles and mansions, notably Craigmillar, having been occupied by her at different times. Her son King James VI was born in the Castle and frequently resided in Holyrood before the Union of the Crowns. Charles I and Charles II are also associated with the Palace and, as is noticed elsewhere, the gallant and ill-fated Bonnie Prince Charlie kept court there in 1745. Other royal personages have since visited and resided in the city, notably George IV, Queen Victoria, Edward VII, and George V.

Since the seventh century when St Cuthbert peram-
bulated the district, preaching Christianity to the Celtic
tribes and the Saxon settlers, and built the cell which

John Knox

preceded the church afterwards dedicated to him, many
famous churchmen and divines have been resident in the
town and county. Three great movements have stirred

deeply the religious life of Scotland—the Reformation, the struggle against Episcopacy, and the Disruption—and prominent churchmen connected with the district were closely associated with all these movements.

John Knox, the greatest of the Scottish Reformers, was born in the neighbouring county of Haddington. When in 1555–56 Knox conducted his first campaign for the Reformed religion, Edinburgh was his headquarters; and later he was appointed minister of St Giles'.

Stone in Parliament Square, Edinburgh, marking approximately the place of Knox's grave in what was formerly the Churchyard of St Giles'

His antagonism to Queen Mary and his famous disputation with her in her palace of Holyrood are recounted by himself in his *History of the Reformation in Scotland.* In the causeway of Parliament Square, formerly part of the churchyard of St Giles', a plain flat stone with brass letters marks the place where, in the words of Regent Morton, "lyeth one who in his life never feared the face of man."

That "fine flower of Scottish theology," Robert Leighton, was a student of Edinburgh University, of which he afterwards became Principal. He was appointed minister of Newbattle in 1641 and there composed the best of his theological writings. These reveal a nature of singular beauty and spirituality.

John Spottiswoode, a native of Midcalder and a historian of the Church of Scotland, rose to be Archbishop of St Andrews and took a leading part in the attempt to impose Episcopacy on the church. He crowned Charles I at Holyrood in 1633.

Another historian and Principal also of Edinburgh University was William Robertson, born at Borthwick in 1721. He became minister of Greyfriars, in the churchyard of which he is buried. His Histories of Scotland, of America, and of the Reign of the Emperor Charles V, notably the last, are dignified in style and sagacious in judgment.

The autobiography of Alexander Carlyle, minister of Inveresk from 1748 to 1805, is specially interesting because of his intimacy with many noted men of his time and of his having been an eye-witness of events in the Forty-five Rebellion and the Porteous Riot. He was a man of imposing appearance and was nicknamed "Jupiter Carlyle." John Home, clergyman and dramatist, author of *Douglas, a Tragedy*, was one of Carlyle's intimates.

Another minister of Old Greyfriars, the Rev. Robert Lee, Professor of Biblical Criticism, was a great innovator in the order of church service and excited much criticism by his introduction of instrumental music.

Prominent leaders in the controversy which ended in the Disruption, and afterwards eminent Free Churchmen, were Thomas Chalmers and Thomas Guthrie, who were alike also in their eloquence as preachers and in their zeal for social reform.

Besides the historians already named, we may note that Sir Archibald Alison, a member of the Scottish bar, who lived at Woodville, Colinton, wrote his *History of Europe* "to prove that Providence was on the side of the Tories"; and that three generations of the Tytlers of Woodhouselee, near Glencorse, all made names as historians. William Tytler vindicated the memory of Queen Mary from the charges brought against her by Robertson and Hume; his eldest son, Alexander Fraser Tytler, afterwards Lord Woodhouselee, was Professor of History in the University and wrote a treatise entitled *Elements of General History*; and Patrick Fraser Tytler, fourth son of Alexander, is well known by his admirable *History of Scotland*.

Another Scottish historian and man of letters, John Hill Burton, lived at Craiglockhart and at Morton House, while Sir John Skelton resided at the Hermitage of Braid near the city.

David Hume was both historian and philosopher. His *History of England* is still read, and his philosophical writings were of considerable importance in the development of modern thought. Other philosophers of the Scottish school associated with Edinburgh were Dugald

Stewart, Sir William Hamilton, and Adam Smith, author of the *Wealth of Nations*. Thomas Carlyle was a student of the University and lived and worked in the city for a number of years. A tablet marks the house at Comely Bank where he stayed after his marriage.

Scott's Statue in the Monument

Of the many literary men associated with the district, Edinburgh's own son, Sir Walter Scott, overshadows the rest as his monument towers over the Princes Street statues. Though the house of his birth in College Wynd has long been swept away, his boyhood's home in

25 George Square, his own first house in George Street, and 39 Castle Street where he resided for 38 years and did most of the work that brought him fame, still remain, and at Lasswade is still to be seen the cottage which was his country home for a few years.

Lockhart, Scott's son-in-law and biographer, was a member of the Edinburgh bar, who failed as an advocate and devoted himself to literature. His *Life of Scott* is one of the greatest biographies ever penned.

Another Edinburgh advocate, James Boswell, wrote the *Life of Samuel Johnson*, admittedly the greatest biography in English. He was the son of a Judge in the Court of Session. Boswell brought his great patron to Edinburgh *en route* for the famous tour in the Highlands and Western Islands, and Dr Johnson was much pleased by the reception he met with from the *literati* of the city.

Attracted by letters of encouragement which the Kilmarnock edition of his poems brought him, Burns came to Edinburgh in 1786 and spent the winter there, lionised by the society of the day. His lodging was first in Baxter's Close, Lawnmarket, and afterwards in St James' Square. While in Edinburgh he erected a tombstone to the memory of Robert Fergusson, his ill-fated predecessor in the art of Scottish poesy, a native of the city, and author of *Braid Claith*, *Leith Races*, and the *Farmer's Ingle*, the last a prototype of the *Cotter's Saturday Night*. Fergusson may be called the poet-laureate of Edinburgh, which he apostrophises as:

> "Auld Reekie! wale o' ilka town
> That Scotland kens beneath the moon."

Burns's Monument and Arthur's Seat

One of the most prominent of Burns's admirers was Henry Mackenzie, the "Man of Feeling," as he was called from one of his writings—sentimental novels of the school of Sterne and Richardson.

Robert Fergusson

Allan Ramsay, author of the *Gentle Shepherd*, carried on business first as a wig-maker and later as a bookseller in the High Street of Edinburgh; and the quaint octagonal house he built for himself on the Castle Hill is now part of the premises of the Outlook Tower facing down upon his statue in the gardens at the foot of the Mound. The house was occupied later by John Galt, the Ayrshire novelist, while writing his *Annals of the Parish*. One of Ramsay's chief services to literature was renewing interest in earlier Scots poetry, notably that of William

Allan Ramsay

Dunbar, who celebrated James IV's marriage in *The Thistle and the Rose*, and who lived about the court in Holyrood for some years before Flodden. A contemporary, Gawain Douglas, son of Archibald Bell-the-Cat, was dean of St Giles' when he wrote his poems, and he

describes local scenery picturesquely in one of the pro-
logues to his translation of the *Aeneid*. Sir David Lynd-
say, poet and satirist, was also a courtier at Holyrood,
where he acted as "keeper" to the infant James V, who
later appointed him Lord Lyon King-at-Arms. His poems

George Buchanan

satirising the faults of the clergy had no small effect in
helping to bring about the Reformation. The greatest
scholar that Scotland has ever produced, George Buchanan,
spent the last years of his life chiefly in Edinburgh, and
was buried in Greyfriars churchyard, 1582.

Three years later William Drummond was born, whose house at Hawthornden was described as "a sweet and solitary seat and very fit and proper for the Muses." There the poet enjoyed the friendship of his literary brethren, notably Drayton, the great Montrose, and Ben Jonson.

At Lasswade and at various lodgings in Edinburgh that erratic genius Thomas de Quincey, "the English Opium-eater," lived from 1828 till his death in 1859, contributing to numerous periodicals his brilliant essays on all manner of subjects. Lasswade School was for three years the scene of the labours of William Tennant, author of *Anster Fair*.

At the end of the Mall in Musselburgh stands a statue to David Macbeth Moir, the beloved physician of the "honest toun" and the "Delta" of *Blackwood's Magazine*. His *Mansie Wauch* is still read for its humorous pictures of Scottish rural life. Mention of *Maga* reminds us of John Wilson, "Christopher North" of the *Noctes Ambrosianae*; William Edmonston Aytoun, his son-in-law, best known by his *Lays of the Scottish Cavaliers*; and Mrs Oliphant, born at Wallyford near Musselburgh, whose *Chronicles of Carlingford*, first published in *Blackwood*, made her reputation as a novelist. Another novelist, Miss Susan Ferrier, a native of Edinburgh, wrote stories characterised by shrewd insight into national peculiarities.

Also a beloved physician was Dr John Brown, whose *Rab and his Friends* and *Pet Marjorie* are enrolled among the immortals.

James Grant, the military novelist, author of the

Romance of War, was born in Edinburgh, where after a short service in the army he resided for many years, and which he loved with a peculiar affection evidenced in his *Old and New Edinburgh* and *Memorials of Edinburgh Castle*. Other historians of the good town were

Dr John Brown

Sir Daniel Wilson and Robert Chambers. Affection for Edinburgh characterised also one of the city's most famous sons, Robert Louis Stevenson, "inheritor of the traditions of Scott with the world pain of his own epoch super-added." Many of Stevenson's writings allude to the

ancient town of his birth and to its romantic neighbourhood, especially Swanston, the little hamlet at the northern
base of the Pentlands, where his father had a country
house, and Colinton, where his grandfather was parish

Robert Louis Stevenson

minister. *St Ives*, *Weir of Hermiston*, *Kidnapped*, *Cat-
riona*, and many of his essays and poems evince intimate
knowledge of the locality and of the idiosyncracies of its
people.

As might be expected, Edinburgh has had many

citizens eminent in knowledge of the law.   Three of
the most famous were contemporaries and friends, Jeffrey,
Brougham, and Cockburn.   Jeffrey is best remembered
by his connection with the *Edinburgh Review*, which
owed its inception to Sydney Smith, then resident in the

"The Cottage," Swanston

city.   Brougham, a man of the most wonderful range of
knowledge and the greatest eccentricity of manners, rose
to be Lord Chancellor of England, and was a protagonist
in the battle for Parliamentary Reform.   Cockburn's
Memorials of his Time is full of interest as a record of
the most brilliant period in the history of Edinburgh,

the first half of the nineteenth century. His home at
Bonaly Tower, beautifully situated on the north slope
of the Pentlands near Colinton, was much resorted to by
his legal and literary friends. Other famous lawyers were

Francis, Lord Jeffrey

the Dalrymples, the Dundases, Forbes of Culloden, and
the witty Harry Erskine.

In medicine the best known name is that of Sir James
Young Simpson, the discoverer of chloroform, who helped
greatly to build up the fame of the University as a medical

school. In physical science Napier of Merchiston, the inventor of logarithms; Hutton and Hugh Miller, geologists; Sir D. Brewster and Professor Forbes, physicists, are all associated with the town or neighbourhood. Hugh Miller worked as a stone-mason at Niddrie; and afterwards, when editor of the *Witness*, he resided in Portobello, where he died in 1856.

Edinburgh has contributed her quota to the ranks of war. Sir David Baird, the hero of Seringapatam and of the well-known tale of "oor Davie," was born in the Lawnmarket. General Dundas, who laid the foundation of British South Africa by his capture of Cape Town, was a member of the famous legal family of that name. Sir Ralph Abercromby, the British general who died in the moment of victory over the French at Aboukir in 1801 and one of the best and bravest of Scottish soldiers, was a citizen of Edinburgh; as was also Sir James Hope, afterwards Earl of Hopetoun, one of "Wellington's Men." The house in George Square where lived Admiral Duncan, the hero of Camperdown, is now occupied as a Ladies' College. He and other naval commanders, Nelson included, admitted their indebtedness to John Clerk of Eldin near Lasswade, who though he had no practical acquaintance with the art of war, evolved a highly successful theory of naval tactics on the ponds of Penicuik estate.

Of many first-rate painters we may note Allan Ramsay, the poet's son, who rose to eminence as portrait painter to George III; Sir Henry Raeburn, a native of Stockbridge and a foundationer of Heriot's Hospital, who

Sir Ralph Abercromby

portrayed all the eminent Scotsmen of his day, and whose reputation is steadily growing; Alexander Nasmyth, his contemporary, the painter of the best known portrait of Burns; John Thomson, minister of Duddingston, the first of Scottish landscape painters and still looked

Sir Henry Raeburn

upon as one of the best; R. Scott Lauder, subject painter and illustrator of Scott; Horatio McCulloch, unrivalled as a delineator of Highland scenery; Sir Noel Paton, poet as well as artist, whose religious, allegorical, and fanciful pictures are so often reproduced; and Sam Bough, an

Englishman who settled in Edinburgh and obtained considerable fame as a self-taught landscapist. Kay, the caricaturist, was a native of Dalkeith. He became a barber and printseller in Edinburgh, whose citizens he immortalised with rude but effective skill. The list may fitly close with Sir David Wilkie, who studied in the Edinburgh school, and whose place in Scottish painting is comparable with that of Scott and Burns in literature.

Mention may also be made of the architects—Adam, who designed the Register House, St George's Church, and the University Buildings; Telford, who planned the beautiful Dean Bridge; and George Meikle Kemp, native of Peeblesshire and draughtsman in Edinburgh, whose monument to Sir Walter Scott is a monument also to his own genius and taste.

25. THE CHIEF TOWNS AND VILLAGES
OF MIDLOTHIAN.

(The figures in brackets after a name give the population in
1911, the asterisk denoting parishes, and those at the end of
each section are references to pages in the text.)

Addiewell is a mining and manufacturing village near
West Calder. It was founded in 1866 in connection with the
oil-shale industry, and produces paraffin-oil, candles, and ammonia.
(p. 77.)

Auchendinny, on Glencorse burn. Near it are Greenlaw
Barracks and Auchendinny House, at one time the residence of
Henry Mackenzie, the "Man of Feeling."

Balerno is a paper-making centre in the Water of Leith
Valley, and a terminus of a branch line of the Caledonian
Railway. (pp. 48, 68.)

Blackhall, on the western outskirts of Edinburgh and close
to the northern base of Corstorphine Hill. Craigcrook Castle
near by was occupied by Constable the publisher, and Francis,
Lord Jeffrey.

Bonnyrigg (2955) is a police burgh with a station on the
Edinburgh to Peebles branch of the North British Railway.
Many of its inhabitants are employed in neighbouring coal mines
and in a local carpet factory. (pp. 64, 69, 160.)

Burdiehouse is four and a half miles south-east of Edinburgh. The original name is believed to have been Bordeaux House, from the French attendants of Queen Mary who settled there. Its lime-works have long been famous. The limestone is interesting geologically as a freshwater deposit. (pp. 54, 65.)

Colinton (6664*) is charmingly situated on the banks of the Water of Leith near a dell which is a favourite resort of city

Craigmillar Castle

holiday makers. With a station on the Balerno branch, it has become a popular suburb and has many pretty villa residences clustered on the sides of the valley. (pp. 65, 68, 99, 156, 161, 168, 177, 179.)

Corstorphine (3870*) stands on the Glasgow road two miles west of Edinburgh. Sheltered from the north and east winds by

Corstorphine Hill and built on a sunny slope it has many attractions as a place of residence. In the eighteenth century the virtues of a sulphurous spring made it a fashionable watering place. "Rest-and-be-thankful" on the east ridge of the hill commands a fine view of Edinburgh and the shores of the Forth. (pp. 22, 30, 63, 101, 103, 110, 120, 156, 161.)

Cousland, an old village near the eastern boundary of the county, was burned by Somerset after the battle of Pinkie. Portland cement is made near by. (p. 65.)

Craigmillar, near Duddingston Loch, is named from the famous old castle. It has sprung up recently on account of the suitability of its water for brewing. Several large malting and brewing firms have erected their plant close to the Suburban Railway. There is also a Creamery and Margarine Factory. (pp. 64, 65, 94, 127, 129, 130.)

Cramond—*v.* p. 38.

Currie (2519*) is one of the paper-making villages in the Water of Leith valley. The river is crossed by an old fourteenth century bridge, which figures in local proverbial lore—"as deep as Currie brig." (pp. 53, 68, 99, 110, 155, 161.)

Dalkeith (7019) is built on the tongue of land between the North and the South Esk just above their junction. It is the market-town for a rich agricultural district and has a Corn Exchange with weekly markets, flour-mills, a foundry, a brewery, and rope, brush, and carpet factories. Dalkeith Palace, residence of the Duke of Buccleuch, is close to the town. It was the head-quarters of General Monk when Governor of Scotland, 1654–1660. (pp. 17, 66, 69, 103, 122, 149, 153, 157, 160, 161.)

East Calder (2873*)—Gael. *Choille-dur*, "a wooded stream"—stands near the banks of the Almond. It has limestone quarries. (p. 103.)

Edinburgh (320,315). Beautiful for situation, the city enjoys natural advantages such as few towns can boast. Built

Princes Street, Edinburgh

upon various ridges and slopes and in the intervening valleys, the town is not only picturesque in appearance but also airy, well-ventilated, easily-drained, and healthy. Its nearness to the sea and to a productive coalfield gives excellent facilities for manufactures and commerce: its educational advantages, its importance as the metropolis of the country, and its historical associations, combined with the beauty of its natural surroundings, make it a favourite place of residence.

Grassmarket, Edinburgh

Edinburgh grew up as a huddle of houses on the long eastern slope or tail of the Castle Rock, on which some sort of fortification must have existed from the earliest times; and walls were built at different dates to enclose and guard the town. The Old Town consisted of the Castle Hill, the Lawnmarket, the High Street, and the Netherbow, with the closes leading off to right and left down the slopes of the ridge. At the other end of the Royal Mile from the Castle stood Holyrood Abbey and under its shelter rose the Burgh of Canongate, a separate town, which extended westwards until it met the Netherbow at the Netherbow Port.

At the time of Flodden the town had burst these narrow bounds and the Flodden Wall had to be built to enclose the southern suburbs. Even as late as 1745 the City proper was a walled town; but after the failure of the walls to withstand the Jacobite attack, they were doomed to destruction as mere obstacles to communication, and now only a few vestiges of them remain.

The narrow limits of the old town necessitated the erection of the tall tenements which were, and still are, so marked a feature of its architecture: and at various periods the need for accommodation has brought about a widening of the city boundaries. That of 1513 has been noticed above. In 1766 George Square with its neighbourhood was designed as an aristocratic southern quarter; and in 1767 the New Town on the north side of the valley which contained the Nor' Loch began to be laid out on the ridge and slope beyond, leading to the Water of Leith. Princes Street occupies the site of an old country road called the Lang Gaitt or the Lang Dykes; and now with its array of varied and handsome buildings, its charming gardens backed by the frowning cliffs of the Castle Rock and the lofty range of the Old Town houses, it is admittedly one of the finest streets in the world.

The New Town contains many notable examples of town-planning and domestic architecture; squares, terraces, crescents, and circuses breaking the monotony of straight streets and giving dignity and spaciousness to those parts of the town. The earth dug out for the foundation of the New Town was piled up in the valley from which the Nor' Loch was drained in 1763. As the Mound it forms a means of communication between the New and the Old Town, as do also the North Bridge and the Waverley Bridge. About the same time the South Bridge carried the line of communication over the parallel valley to the southern suburbs, which now extend beyond the ridge of the Boroughmuir to the gentle slope towards the valley of the Pow or Jordan Burn at

Morningside and Newington and even to the rise beyond. Great extensions have also taken place to the east towards Portobello and to the west towards Slateford and Corstorphine.

Edinburgh is remarkably well supplied with open spaces, and the extent of the town is therefore much greater than the number of its population might lead one to suppose. Princes Street Gardens, Queen Street Gardens, the Royal Botanic Garden and the Arboretum, the Meadows and the Links, the King's Park,

Old Quadrangle, Edinburgh University

the Braids and Blackford Hill, and the various city parks along with numerous private grounds make up a very considerable portion of the city area.

The old University occupied the site of the Church of St Mary in the Fields, the scene of Darnley's murder in 1567; and was founded under a charter of James VI (1582). The present edifice was begun in 1789 and portions of the quadrangle were built at various times to the designs of Adam and Playfair. The main front faces on to South Bridge Street and the opening up of Chambers Street has given opportunity to improve the northern aspect. The general style of the architecture is Graeco-Italian: on both the external and the interior view of the quadrangle the buildings are dignified and impressive and the lofty dome over the entrance, surmounted by the gilded figure of Youth

Fettes College

as a torch-bearer, is a conspicuous feature in a distant view of the city.

The Medical School now occupies a separate pile of buildings, opened in 1880, in Teviot Place and Meadow Walk, where the classrooms, arranged around two quadrangles, are thoroughly equipped for instructional purposes. Adjoining the New Buildings are the Music Classroom, the Students' Union, and the University Hall, the last gifted by Mr William McEwan, formerly M.P. for one of the city divisions. A new Engineering department has also been set up in Infirmary Street.

Close to the University in Chambers Street is the Royal Scottish Museum, a handsome building in Venetian Renaissance style, the foundation stone of which was laid by Prince Albert in 1861. The Museum has a fine Natural History collection, an industrial department with specimens of raw materials and manufactured products in various stages illustrating all the more important industries of the country, and Egyptian, geological, educational, and other sections.

Important as affording ample scope for clinical teaching to students of the University, and notable as a monument of private charity, the Royal Infirmary in Lauriston deserves special mention.

Unpardonable blunders were committed by the authorities of Edinburgh when they permitted the Nor' Loch to be drained, the artificial Mound to be heaped up in the dried basin, and the railway to be constructed through the valley. But the best has been done to make up for these mistakes. The Mound has been dignified by the erection upon it of the fine Doric temple long known as the Royal Institution, with its colossal seated statue of Queen Victoria, and the sister building in Ionic style, the National Gallery. The sides of the valley have been beautifully laid out in gardens and along the Northern esplanade has been placed a line of monuments, chief among them being the lofty Gothic spire canopying the seated figure of Sir Walter Scott and having clustered in its niches numerous statuettes representing characters, historical and imaginary, portrayed in his works. Ascending the Mound to reach the Old Town we have on our left the Bank of Scotland, on our right New College—the United Free Church Divinity Hall.

Of the schools whose fame has won for Edinburgh a great reputation as a seat of learning, the oldest is the Royal High School, which as the Grammar School of the town was founded in

1519 and after being located on various sites now occupies a picturesque building, mainly Doric in style, upon a terrace on the southern face of the Calton Hill. Crowning the hill immediately above it, rise the twelve finely fluted and beautifully proportioned columns of the unfinished National Monument, projected in 1816 and designed by W. H. Playfair as a copy of the Parthenon at Athens to commemorate the Scottish heroes who fell in the

Royal High School, Edinburgh

Napoleonic wars. Nelson's monument close by, though far from picturesque, is a prominent feature in views of the city and is made use of to display on its summit a huge time-ball, which is raised daily to drop exactly at one o'clock, when also a time-gun is fired at the Castle. The gloomy castellated building opposite the High School is the Calton Jail.

Parliament Square, behind St Giles', is surrounded by a range of classical buildings with an arcaded piazza. Here are housed the Law Courts H.M. Exchequer Office, and the libraries of certain legal fraternities, including the Advocates' Library, one of five in the United Kingdom possessing the right to a copy of every book published in this country. Near at hand are the Midlothian County Buildings; and on the opposite side of the High Street the Municipal or City Chambers are ranged round three sides of a quadrangle, while on the fourth or southern side is a screen of

Greyfriars Church, Edinburgh

seven archways covered over by a platform with ornamental balustrade and vases. Within are public offices and the Council Chamber where the "Fathers of the City" meet in conclave.

In Greyfriars Churchyard took place the signing of the National Covenant in 1638, the momentous step in the rebellion against Charles I's ecclesiastical policy. Here also many of the martyred Covenanters were buried.

On the Castle Hill stands a handsome Gothic building with a lofty spire, the meeting place of the General Assembly—the highest Court—of the Church of Scotland. (pp. 4, 5, 7, 8, 10, 11, 15, 17, 20, 37, 46, 48, 50, 53, 54, 57, 58, 61, 63, 64, 65, 66, 67, 68, 69, 70, 71, 81, 87, 94, 97, 99, 101, 102, 103, 105, 108, 110, 112, 122, 127, 128, 134, 140, 141, 153, 156, 157, 159, 163, 164, 166, 167, 168, 170, 172, 174, 176, 180, 183.)

Gilmerton occupies a ridge, 400 feet above sea-level, four miles south-east of Edinburgh. Gilmerton carters, who at one time supplied the city with coal and lime, were a numerous body, whose "Play" was an annual event of note in the locality. A cave hewn out of the rock by a blacksmith, George Paterson, who lived in it with his family about the beginning of the eighteenth century, used to be considered a great curiosity. Coal and iron mines are worked in the neighbourhood. (pp. 64, 74, 154.)

Gorebridge is situated in the heart of the Midlothian Coalfield, near the collieries of Arniston, Dalhousie, Newbattle, and Vogrie. At Stobs Mill the first gunpowder factory in Scotland was opened in 1793. (pp. 19, 75.)

Granton, a busy port on the Firth of Forth three miles north-north-west of Edinburgh, of which it is now a part. (pp. 4, 37, 40, 41, 53, 70, 79, 84, 88, 90, 156.)

Inveresk (20,360*)—Gael. *Inbhir-uisge*, "confluence of the water"—occupies a slope leading up to the lofty right bank of the Esk about half a mile from its mouth. Its pleasant situation and healthy climate have attracted residents ever since Roman times; and it has many picturesque old-fashioned houses. The church, on the site of the Roman Praetorium, is the successor of one built soon after the introduction of Christianity, which was preached in by George Wishart just before his martyrdom. (pp. 96, 99, 108, 109, 161, 167.)

Juniper Green, on the high left bank of the Water of Leith, five and a half miles south-west of Edinburgh, contains many villas of Edinburgh business people. A paper-mill, a flour-mill, and two snuff-mills give employment to the inhabitants, and there are two golf courses in the neighbourhood. (pp. 68, 70, 107.)

Sir Walter Scott's Cottage, Lasswade

Lasswade (880), a police burgh since 1866, is a quaintly irregular village scattered along the hollow and the steep banks of the Esk, here crossed by a bridge. Its romantic situation supplied some of the details in the description of Gandercleugh in the *Tales of my Landlord*. The thatched cottage in which Scott lived from 1798 to 1804, and where he was visited by Wordsworth, is still standing; Tennant, author of *Anster Fair*, was schoolmaster from 1816 to 1819; Drummond of Hawthornden is buried in one of the aisles of the old church, where a monument has been erected to his memory; from 1840 to 1859 De Quincey occupied Man's Bush Cottage, now De Quincey Villa, near Polton.

Most of the work people are employed in the paper-mills, the first of which was opened in 1750. (pp. 17, 68, 71, 110, 149, 160, 161, 170, 175, 180.)

Council Chambers, Leith

Leith (80,489) is the sixth largest town in Scotland. The long street, Leith Walk, which connects it to Edinburgh, was originally a line of earthwork built in 1650 when Cromwell invaded Scotland. Owing to town improvements very little of old Leith now remains though it has many interesting historical associations. It was even at one time a walled town, its fortifications having been built by French supporters of Mary of Lorraine, who became Queen Regent in 1554. The Links still show traces of the earthworks thrown up by the opposing Protestant party, when the town was besieged in 1559-60. In 1650 Cromwell built the citadel of Leith, long since destroyed.

Besides the work connected with its important harbour, various industries are carried on, as saw-milling, flour-milling, engineering, brewing, distilling, manufacture of cattle-feeding stuffs and chemicals, rope-, twine-, and sail-making, and coopering. (pp. 4, 15, 22, 37, 41, 48, 66, 68, 69, 70, 79, 80, 81, 83, 84, 85, 87, 88, 94, 99, 103, 156, 159, 162.)

Liberton (8360*) was called the "leper town" from a hospital which once stood at Upper Liberton. The village is built on the top of a ridge 356 feet above sea-level about two and a half miles south-east of Edinburgh, and the parish church with its four-pinnacled tower is a conspicuous object in the landscape. Reuben Butler in the *Heart of Midlothian* was schoolmaster of Liberton, and a fine old baronial mansion, Peffer Mill, in the vicinity was the prototype of "Dumbiedykes." The nuns of the Sciennes were accustomed to make an annual procession to the well of St Catherine at Liberton, a bituminous spring formerly much resorted to by persons suffering from skin diseases. The stern Puritans of Cromwell's army destroyed the shrine as a place of idolatry. The eastern end of the ridge on which Liberton stands is occupied by Craigmillar Castle. (pp. 53, 125, 161.)

Loanhead (3483) a police burgh since 1884, is a mining and manufacturing town half a mile from the left bank of the North Esk. It has a station on the Roslin and Glencorse branch of the North British Railway. The inhabitants are mostly engaged in mining and paper-making. (pp. 69, 74, 76, 160.)

Midcalder (3249*) is situated on the south bank of Almond Water near the junction with Linhouse and Murieston Burns. Employment is obtained in the oil, paper, and gunpowder works;

Polton

and fairs are held twice a year, Ca'ther Fair at one time having been an important fixture for the surrounding district. The Parish Church was founded early in the thirteenth century. Calder House, the seat of Lord Torphichen, is a very old building with walls seven feet thick. Spottiswoode, Archbishop of St Andrews and historian of the Church of Scotland, was a native of the village. (pp. 14, 69, 122, 155, 161, 167.)

Musselburgh (15,938), the terminus of a branch of the North British Railway, six miles east from Edinburgh, is a parliamentary burgh situated mainly on the right bank of the Esk but having various outlying suburbs—Fisherrow on the opposite bank of the river; Magdalen Bridge at the mouth of Burdiehouse Burn, a mile to the west; Newbigging, an old village of one long street on the south side of the town; and Levenhall at the east end of the Links. The Links have long been famous for archery, golf, and horse-racing. At the town end of the Links once stood an ancient chapel and a hermitage dedicated to Our Lady of Loretto. Near the site is a boarding school which is conducted on English lines and which has attained a considerable reputation. The villa of Loretto was once occupied by Lord Clive. Much of the land round the town is used for market gardens, for which its fertility makes it specially suitable. A good many of the inhabitants work in neighbouring coal-pits. There are also paper-works, breweries, oil-crushing works, net, rope, and wire factories, and flour mills. Musselburgh was named "honest toun" from its motto—"Honestas." (pp. 19, 37, 43, 66, 68, 69, 70, 81, 83, 84, 96, 99, 106, 109, 134, 151, 153, 156, 157, 159, 162, 175.)

New Craighall is a large mining village on the boundary of Liberton and Inveresk parishes with a station at Newhailes on the Musselburgh branch of the North British Railway.

Newhaven, a fishing village on the Firth of Forth, two miles north of Edinburgh. (pp. 4, 40, 41, 79, 81.)

Newtongrange is an important mining centre in Newbattle parish, two miles south of Dalkeith.   (pp. 19, 75.)

Oakbank, in Kirknewton parish near Midcalder, has oil-refineries.   (p. 77.)

Penicuik Church

Pathhead, five miles south-east of Dalkeith, takes its name from its position on the summit of an ascent from the River Tyne, which is here crossed by a fine bridge of five arches each 80 feet high and 50 feet in span.   The ancient village of Ford lies on the opposite side of the river.   The neighbourhood, including the grounds of Oxenfoord Castle and Vogrie House, is finely wooded and picturesque.

Penicuik (2736)—Celtic *pen-y-côg*, "the hill of the cuckoo"—stands 600 feet above sea-level on the left bank of the North Esk ten miles south of Edinburgh. It is the terminus of a branch line of the North British Railway, and is also served by Pomathorn and Glencorse stations on other branches. It has a suburb, Kirkhill, with a population of 517. Penicuik depends chiefly on the paper-making industry though it has also an iron foundry. Many of the employees of neighbouring collieries reside in the town. Paper-making was begun here about 1709 and there are at present three mills Valleyfield, Low, and Bank Mills, belonging to Messrs Cowan & Sons; and one, Eskmill, belonging to Messrs Brown & Co. In the grounds of Valleyfield a monument commemorates the connection of the town with the Napoleonic wars. From 1810 to 1814 Valleyfield Mill and Eskmill were occupied as barracks for 6000 French prisoners of war, 300 of whom died, and were buried near where the monument stands. The vicinity of the town presents a wonderful variety of landscape from the bare slopes of Carnethy and Scald Law to the deep thickly wooded gorge of the Esk and the well-cultivated plains to the north and east. Penicuik House, a fine Grecian building erected in 1761, was burned in 1899 and now stands a roofless ruin in its picturesquely romantic grounds. An obelisk on a height on the estate commemorates the friendship of its owners with Allan Ramsay, the scenes of whose pastoral play, *The Gentle Shepherd*, are laid in the neighbourhood. (pp. 28, 35, 64, 68, 69, 76, 103, 160, 161, 180.)

Portobello is now included municipally as one of the wards of Edinburgh. Its fine stretch of level sand has long made the town a favourite resort. A valuable bed of clay has given rise to thriving brick, tile, and earthenware works; and the making of glass bottles is also carried on. The railway station is an important junction with numerous sidings for the shunting and marshalling of trains. (pp. 37, 41, 42, 54, 64, 69, 70, 109, 153, 159, 162, 180.)

Valleyfield Paper Mills and Penicuik

Pumpherston in Kirknewton parish, is a village of recent growth and is engaged in the oil-shale industry. It produces the largest annual output of sulphate of ammonia in the kingdom. (p. 77.)

Ratho (1882*) is an ancient village about nine miles south-west of Edinburgh. It consists mainly of one long street on a hill rising to 320 feet above sea-level and sloping steeply to the east. The Union Canal passes close to the foot of the village. Market gardens and stone quarries give employment to the inhabitants. The place boasts two poets; Joseph Mitchell "the poet of Ratho," who published a book of poems in 1724; and the Rev. William Wilkie, parish minister (1752–1761) and afterwards Professor of Natural Philosophy, St Andrews, whose *Epigoniad* gained him, undeservedly, the name of the Scottish Homer. (pp. 13, 31, 53, 63, 64, 103, 105, 161.)

Restalrig is an old village in the parish of South Leith. Up to the time of the Reformation it was the parish town and its church, dedicated to St Triduana, was long a place of pilgrimage. The fine old Chapter House has recently been restored. The barony of Restalrig was twice forfeited, first by Sir Robert Logan for his share in the Gowrie Conspiracy, and again by Lord Balmerino, who was "out" in the 'Forty-five. (pp. 20, 22, 69, 105, 122.)

Rosewell, four miles south-west of Dalkeith, is inhabited chiefly by colliers. (p. 74.)

Roslin is about six and a half miles south-west of Edinburgh and has a station on the Glencorse branch of the North British Railway. Under the protection of the powerful family of St Clair it was once a place of considerable importance. The fame of its chapel and castle, and the wild beauty of the glen, through which the North Esk flows at this part of its course, attract many visitors. A gunpowder factory, a carpet factory, paper-mills, and coal pits supply employment. (pp. 17, 53, 69, 76, 92, 94, 103, 105, 118, 130.)

Slateford is notable as having close together a bridge over the Water of Leith, a railway viaduct for the Balerno branch of the Caledonian Railway, and an aqueduct 65 feet high and 500 feet long carrying the Union Canal over the river. A bleachfield and dyeworks employ many of the inhabitants. The new slaughterhouses and cattle markets of the Edinburgh Corporation cover an extensive area near the railway station. (pp. 32, 68, 69 101, 156.)

Stow (1317*) stands at a height of 580 feet, on the left bank of Gala Water, 24 miles south-east of Edinburgh, and is sheltered all round by hills. Like other towns in the basin of the Tweed, Stow is engaged in the making of woollen cloths or tweeds. The parish church is a fine example of modern ecclesiastical architecture. (pp. 54, 69, 110, 161.)

Straiton has somewhat decayed in recent years since the closing of its oil-shale works. The Burdiehouse Limestone is still worked. (pp. 65, 77.)

Temple (363*) was so called from being at one time the chief seat in Scotland of the Order of Knights Templars. The old church was part of their preceptory. The village, 605 feet above sea-level, is on the right bank of the South Esk about seven miles south-west of Dalkeith. (p. 161.)

Wallyford, on account of recent development in the mining industry, has grown to be quite a populous place. It lies on the eastern edge of the coalfield about a mile and a half from Musselburgh. (pp. 75, 76, 175.)

West Calder (7717*) has grown greatly since the exploitation of the oil-shale and coal seams of the district, begun in 1861. Mossend, Gavieside, and Addiewell are all in the immediate neighbourhood, and owe their rise to the same cause. West Calder has a station on the Edinburgh and Glasgow section of the Caledonian Railway, 16 miles from Edinburgh and 31¼ from Glasgow. (pp. 77, 161.)

Fig. 1. Comparative areas of
Midlothian and all Scot-
land

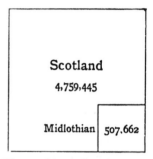

Fig. 2. Comparison in population of Midlothian and
all Scotland

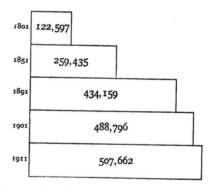

Fig. 3. Growth of population in Midlothian

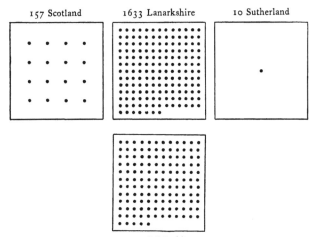

157 Scotland   1633 Lanarkshire   10 Sutherland

1373 Midlothian

Fig. 4. Density of population of all Scotland, Lanarkshire,
Sutherland, Midlothian

(*Each dot represents 10 persons*)

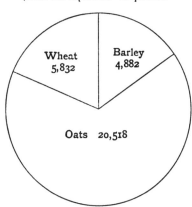

Wheat
5,832

Barley
4,882

Oats   20,518

Fig. 5. Comparative areas under cereals in Midlothian

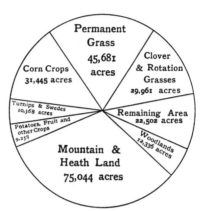

Fig. 6. Comparative areas under various crops in Midlothian

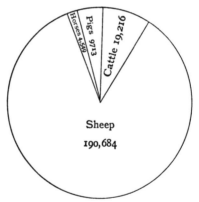

Fig. 7. Comparative numbers of different kinds of live stock in Midlothian in 1909

www.ingramcontent.com/pod-product-compliance
Ingram Content Group UK Ltd.
Pitfield, Milton Keynes, MK11 3LW, UK
UKHW042143280225
455719UK00001B/57

9 781107 620810